CONVERSATIONS

ON

CHEMISTRY.

IN WHICH

THE ELEMENTS OF THAT SCIENCE

ARE

FAMILIARLY EXPLAINED

AND

ILLUSTRATED BY EXPERIMENTS.

IN TWO VOLUMES.

VOL. I.
ON SIMPLE BODIES.

LONDON:
PRINTED FOR LONGMAN, HURST, REES, AND ORME,
PATERNOSTER ROW.

1806.

Frontispiece: Title page to the First Edition

CHEMISTRY IN THE SCHOOLROOM: 1806

selections from

Mrs. Marcet's *Conversations on Chemistry*

chosen and edited by Hazel Rossotti

Senior Research Fellow, St. Anne's College, Oxford, U.K.

1663 LIBERTY DRIVE, SUITE 200
BLOOMINGTON, INDIANA 47403
(800) 839-8640
www.AuthorHouse.com

© 2006 Hazel Rossotti. All Rights Reserved.

No part of this book may be reproduced, stored in a retrieval system, or transmitted by any means without the written permission of the author.

First published by AuthorHouse April. 06, 2006

ISBN: 1-4259-0534-X (sc)

Printed in the United States of America
Bloomington, Indiana

This book is printed on acid-free paper.

Jane Marcet seems to have been fortunate in her offspring, her offspring-in-law, and her husband, who doubtless knew more chemistry than she did. So it seems appropriate that this book be dedicated to:

Francis
Heather
Ian and
Nicole

the order being strictly alphabetical.

EXPLANATIONS AND THANKS

Chemistry in the Schoolroom is a selection from the first edition (London, 1806) of Jane Marcet's *Conversations on Chemistry*. Her own Preface, in which she explains why she wrote it, is given in full (see p. xxii), and readers are strongly recommended to look at this before dipping into the rest of the book. In my Introduction, I have tried to sketch out enough of the biographical, historical and literary background of Jane Marcet's work to allow it to be enjoyed in its context.

The selection itself is personal. I first sought out the book from idle curiosity; I had come across a few (but always the same) quotations from it in my childhood, but only recently had I heard the name of its author. One cursory glance at the two small volumes suggested that I should be depriving myself of a lot of pleasure if I only dipped into them. So I read them in full. Naturally, some topics seem to sparkle more than others. Tastes and priorities have doubtless changed during the two centuries since the book was written; but maybe Jane Marcet's youthful readers never did find the various categories of caloric as much fun as flames, bangs and changes of colour, or as interesting as the many ways in which chemistry enriches daily life. Since I have no reason to suppose that my tastes are perverse, I have sought to share my enjoyment by selecting those passages which gave the most pleasure to me. The great variety of topics discussed in these excerpts arose from no conscious selection on my part, but reflects the wide range of Jane Marcet's own interests and makes for a pleasingly balanced collection.

The selections follow Jane Marcet's order; and her own titles are used for each *Conversation* and, wherever possible, for the excerpts. Where necessary, I have inserted section titles which I hope are in keeping with the text. The original page numbers of each excerpt are given in a footnote, so that those readers who are lucky enough to have access to the first edition or to its facsimile may trace it if they so wish. I have tried to be modest with other footnotes, and have left it to the reader to relate Jane Marcet's views on chemistry to the received wisdom of our own day. After much consideration, I decided to stick to the original spelling (which is sometimes inconsistent) and largely to the original punctuation.

Many people have helped me to prepare this book. Indeed, they have not merely helped, they have been invaluable. My pleasure in thanking them is tempered by anxiety about possible omissions, which I hope will be viewed with that generous understanding which accompanied the original help. Librarians in Oxford seem to be a particularly saintly breed, both in my own college, St. Anne's, and in all the branches of the University's Library Service with which I have had contact, mainly in the Radcliffe Science Library, the Bodleian Library and the Imaging Service. And at the other end of Oxford, the germ of IT-literacy was implanted in me and nurtured with great skill and patience by Reg Cox, the computer manager at Wood Farm Community Centre.

I am also greatly in debt to those who, purely as individuals, have been most generous with their time and expertise. In particular, I must thank Kathryn Sutherland, whose dinner-time query as to whether I knew of Jane Marcet was origin of my renewed interest in *Conversations on Chemistry*. She and David Knight have both been kind enough to read my *Introduction* and to make comments which have much improved it (although naturally the remaining infelicities of commission, omission and style are all my own).

I much enjoyed meeting Bette Polkinghorn and hearing of some of the experiences she had whilst she was working on her most valuable biography of Jane Marcet. A large number of people have helped me in a wide variety of ways. Some have suggested colleagues whom I might contact and literature which I should consult; others have sent me their own unpublished work, or offers of hospitality. Long-suffering IT experts have been generous with their time and patience. Not one of my requests for help went unanswered, whether by letter, e-mail or phone, even from overseas. And to all this help many have added the elixir of their encouragement. Amongst those to whom I am indebted are: Ian Burnell, Michael Bott, Janet Browne, Katherine Cotter, Honor Farrell, Pauline Heath, Ele Hunter, John Issitt, Edgar Jenkins, Elizabeth Morse, Julia Phillips, Richard Pring, Julia Saunders, Sheila Smith, Chris Stray and John Whiteley. I much fear that names of others to whom I am no less grateful will come to mind as soon as it is too late to include them, and for this I apologise in anticipation.

It has been a privilege to have been helped by two of Jane Marcet's great, great, great grandsons, Tom, and the late David, Pasteur. Tom most kindly lent me a transparency of her portrait for reproduction on the front cover, while David gave me a copy of one of her letters, including her signature which is also incorporated in the cover design. I am greatly indebted to them both. My colleague, Marjorie Reeves, was a great source of inspiration, encouragement and practical advice to within a few weeks of her death at the age of 98; such was her enthusiasm that I felt able to visit her only if I had made appreciable progress since our previous meeting.

Members of the AuthorHouse team (including Jo Barber, Ashley Eller, Akram Ibrahim, Peter Voakes, and Emma Williamson) have converted electronic information into the reality of a book with kindly discipline and minimal fuss.

It is usual for authors to thank members of the family, often for both their forbearance and their support; my husband, Francis, despite his comments on "that woman of yours" has in general been not only forbearing, but also a fount of ideas and sound advice. His lynx-eyed reading of the work has been invaluable; and those errors which are now present must surely be late interlopers. Our daughter Heather has also been most generous with her time, and with her IT expertise. This book owes a great deal to them both and I am indeed very grateful.

Hazel Rossotti
St. Anne's College
Oxford OX2 6HS
UK

2005

LIST OF CONTENTS

Explanation and Thanks

List of Contents

List of Illustrations

Background

Introduction   i

Jane Marcet's Preface   xxii

Selections from Vol. I, *On Simple Bodies*

(Capital Roman numerals refer to Conversations, with their original titles and spelling. The titles of the individual selections have been added by the editor).

   I.   **On the General Principles of Chemistry**   1
        The scope of chemistry, 1   Simple or elementary bodies, 5   Chemical attraction, 7   Attraction or affinity?, 9   Decomposition, 10

   II.   **On Heat and Light**   12
        The separation of heat from light, 12   Caloric and heat, 15   Thermometers and temperature scales, 18

   III.   **Continuation of the Subject**   22
        "Hot" and "cold" objects, 22

   IV.   **On Specific Heat, Chemical Heat and Latent Heat**   24
        Steam heating, 24   Is caloric a substance?, 26

V. **On Oxygen and Nitrogen**   28
   Combustion of wood and coal, 28 Prerequisites for combustion, 31 The separation of nitrogen from air, 33 The collection of nitrogen and the combustion of iron, 34 Exhaled air, 37 The insignificance of nitrogen, 37

VI. **On Hydrogen**   39
   Hydrogen gas as generator of water, 39 Water as an oxyd, 41 The decomposition of water, 42 The combustion of hydrogen gas, 43 Hydrogen gas in soap bubbles, 48

VII. **On Sulphur and Phosphorus**   53
   The sublimation of sulphur, 53 The combustion of sulphur, 56 Acids, 57 Sulphurated hydrogen gas, 59 The discovery of phosphorus, 60 The combustion of phosphorus, 62 Phosphorus and sulphur, 64 Phosphorated hydrogen gas, 64

VIII. **On Carbone**   67
   The imitation of nature, 67 Carbonic acid, 68 Hot coals and water, 71 Candles and oil lamps, 72

IX. **On Metals**   75
   Metals, 75 The combustion of iron filings, 77 The decomposition of water by metals, 78 Metals and acids, 80 Crystallization, 84 Alloys, 86 Sympathetic ink, 87

X. **On Alkalies**   89
   Alkalies and earths, 89 Ash and soap, 90 Carbonat of Potash 91 Glass, 93 Ammonia, 94

XI. **On Earths**   97
   Gemstones, 97 Silex, 98 Alumine, 99 Lime-water and carbonic acid gas, 100

Selections from Vol. II, *On Compound Bodies*

XII. On the Attraction of Composition   103
The laws of chemical attraction, 103

XIII. On Compound Bodies   106
The classification of acids, 106

XIV. On the Combinations of Oxygen with Sulphur and with Phosphorus; and of the Sulphats and Phosphats   110
Concentrated sulphuric acid, 110   Antidotes to acid poisoning, 113   Sulphurous acid as a bleaching agent, 114   Sulphats of potash and soda, 115

XV. On the Combinations of Oxygen with Nitrogen and Carbone; and of Nitrats and Carbonats   117
Nitric acid, 117   Aurora borealis, 119   "Exhilarating gas", 120   Gunpowder, 122   Carbonic acid gas, 124

XVI. On the Muriatic and Oxygenated Muriatic Acids; and on Muriats   128
Oxy-muriatic acid, 128   Muriat of ammonia, 129   Oxy- muriat of potash, 130

XVII. On the Nature and Composition of Vegetables   133
Organised bodies, 133   Sugar, 135   Camphor, 137   Tannin, 137

XVIII. On the Decomposition of Vegetables   139
The combustion of alcohol, 139

XIX. History of Vegetation   143
Vegetation around London, 143   Agriculture and manufacture, 143   Water, photosynthesis and respiration, 144

XX. On the Composition of Animals   146
Elements in nature, 146   The diversity of chemistry, 147   Prussic acid, 147

XXI.	On the Animal Economy	149
	Bodily exercise, 149	
XXII.	On Animalisation, Respiration and Nutrition	150
	Respiration, 150	
XXIII.	On Animal Heat; and on Various Animal Products	153
	Body temperature, 153 Curds and whey, 155 *Finis, 155*	

Postscripts

Appendices 157
I. A contemporary review, 157   II. Michael Faraday's appreciation, 158

Index 160

About the editor 167

LIST OF ILLUSTRATIONS

Frontispiece: Title page to first Edition

Plate I, p. 18

Plate II, p. 21

Plate V, p. 32

Plate VI, p. 44

Plate VII, p. 50

Plate VIII, p. 54

Plate IX, p. 79

Plate X, p. 141

All the illustrations within the book are reproduced by permission of The Bodleian Library, University of Oxford from *Conversations on Chemistry: in which the elements of that science are familiarly explained....* by Mrs. Marcet. 1806. [Reference (shelfmark) 1933 e.572]. The images were provided by the Oxford University Libraries Imaging Services. The plate numbers used in the present selection are those in the original work.

The design for the cover incorporates a portrait of Jane Marcet which Tom Pasteur kindly allowed me to use. John Whiteley, of the Ashmolean Museum in Oxford, drew my attention to the red book at the bottom left-hand corner of a transparency of the painting; on it we can just make out the year as 1834 (or perhaps 1839) and a very unclear signature. On the painting, Tom Pasteur has deciphered this as *J. Hornung*, who is presumably Joseph Hornung (1792-1870), a Swiss artist and teacher who painted portraits and landscapes.

INTRODUCTION

Conversations on Chemistry was published almost two hundred years ago and is arguably still one of the foremost works in the field which we now call Public Understanding of Science. Its two pocket-sized duodecimo volumes, with only 13 x 7.5 sq cm of text per page, had a profound influence on the teaching of chemistry on both sides of the Atlantic. The reasons for writing the book, and for presenting it as a series of dialogues, are clearly set out in the author's *Preface* which is reproduced here in full[1]. The name of the author was not revealed until over a quarter of a century later; but her sex was freely admitted, as was her status as a very recent learner of chemistry. Eventually, in 1832, the name *Mrs. Marcet* appeared on the title page of the twelfth edition.

Born in London in 1769, Jane Marcet was the wife of Alexander (or Alexandre) Marcet, a Swiss-born physician who worked in London.[2] Her father[3], Antoine François Haldimand, was also of Swiss origin, although he had been born in Turin, and had lived there until he came to London on an exchange visit. Aged twenty, he worked briefly for Joshua Pickersgill, a silk-manufacturer whose son was sent to the Haldimand family in Italy. Young Haldimand taught his mentor the new Italian accounting techniques; and he also made money from speculation in silk. He later amassed considerable wealth from property development and banking, attributing his success to his total integrity. In 1768, Haldimand married Joshua Pickersgill's daughter and moved to Clapham, which was then an

---

[1] See p.xxii.
[2] Many of the biographical details, from the Marcet family papers, are given in (i) B. Polkinghorn, *Jane Marcet, An Uncommon Woman*, Forestwood Publications, Aldermaston, UK (1993). Shorter, more recent accounts may be found in (ii) Morse, E., in *Oxford Dictionary of National Biography*, New Edition, ed. Matthew, H.C.G., and Harrison, B., Oxford University Press (2004) and (iii) Fyfe, A., in Jane Marcet, *Conversations on Chemistry*, Thoemmes Continuum, Bristol, UK and Edition Synapse, Tokyo (2004). This work is a facsimile of the first (1806) edition, from which the present excerpts are taken.
[3] See Farrell, H., *Gentlewomen of Science – The Role of Women in the London Scientific Circle 1800-1875*, M.Phil., microfiche, University of Leeds (1994) for genealogical, and much other, material.

outlying village, favoured by privileged Londoners. Here, in 1769, their first child was born, and named Jane after her mother. She was to live until 1858; but of her younger siblings, five died in infancy, and another two before reaching maturity. By the time Jane herself was adult, there remained only one sister (Sarah, who was three years her junior) and three yet younger brothers of whom one (William) became a Director of the Bank of England.

At that time, it was usual for young children in well-to-do English families to be looked after by nursemaids. When older, boys were sent to school, where they often had a rigorous diet of classical languages, while many of the girls stayed at home and learned ladylike accomplishments such as music, dancing and painting from a governess.[1] But Jane's father had progressive ideas and wanted his daughters and his sons to have the new type of education being advocated by reformers. Both girls and boys should be taught the exciting new ideas in science by observing natural phenomena and doing simple experiments. Some extremists even thought that science was a subject particularly suitable for girls, as it combined intellectual activity with domestic practicalities. The Haldimand girls indeed had a governess, but they were also taught maths, astronomy and philosophy (much of which we should now call "science")[2] by tutors who visited their house. Jane did attend school for two months but is said to have found the lessons very boring compared with those she had at home.

When Jane was just fifteen, her mother died of complications following childbirth and as Jane was the eldest daughter, she became mistress of the house. She was now in charge of the housekeeping and the servants, and also responsible for overseeing the education of her younger siblings. Once or twice a week she had to act as hostess at her father's lavish parties. The forty-or-so dinner guests included notable politicians, writers, scientists, persons of fashion and important visitors to London; after dinner, yet more guests arrived to join them for dessert. The conversation, which was often so interesting that some of the guests asked if next time they might bring their friends, would surely have stimulated the intellectual curiosity of an intelligent teenager like Jane. Her education was further extended by foreign travel: to Switzerland to meet her relatives, and, with her father and sister, to France and Italy, where she developed an interest in painting.

In 1795, one of the friends brought by a regular guest was Alexander Marcet, who got on well with both Jane and her father; but he soon left London to study medicine in Edinburgh. When he returned in 1797, his visits to the Haldimands were resumed, as was his friendship with Jane.

---

[1] See p. xx, note 3.
[2] See p. xx, note 4.

The couple became engaged in the autumn of 1799, and were married in December. At that time, it was unusual for a woman to be as old as thirty when she married, and also to be older than her husband, albeit in this case only slightly so.

The Marcets chose to live with Jane's father and his three sons; his younger daughter Sarah had married and left home some years earlier. The house in Clapham was renovated to accommodate the combined household, which now numbered about thirty; and it was still supervised by Jane Marcet, who also continued to act as her father's hostess. Alexander Marcet worked as a physician at Guy's Hospital, London, and did research both in medicine and in chemistry, on which he also gave lectures. He had some notable scientific friends, such as Edward Jenner, who introduced vaccination against smallpox, and William Hyde Wollaston, a physician who, like himself, had a chemical laboratory in his home and eventually abandoned medicine for chemistry. Such people were now added to Antoine Haldimand's guest-list. One of Alexander's acquaintances was Humphry Davy[1], Lecturer in Chemistry at the Royal Institution in London. Davy's lectures were a great draw for the fashionable and intellectuals alike, and were open to women as well as to men. At that time chemistry was developing very fast, with many discoveries, such as the isolation of sodium and potassium, being made by Davy himself. He gave excellent lectures, illustrated by experiments which worked; and he was remarkably handsome.

Jane Marcet joined the audience, not because she already had a special interest in chemistry, but because she wanted to learn more about the chemical discoveries which were so frequently discussed by many of her guests. But perhaps the lectures were of the type where a charismatic lecturer carries his audience with him, and only later do its more thoughtful members realise that they do not understand all that was said as well as they thought they did at the time. Jane Marcet, at any rate, felt the need to seek further discussion and explanation from her husband[2]. She soon became hooked by the subject, studied it beyond the lectures and learned practical techniques from her husband in his lab at their home. With their scientific friends she was able to discuss new experimental results and their unpublished hypothetical explanations. It was her husband and their friends[3] who encouraged her tentative idea of writing a chemistry book. As

---

[1] See Knight, D., *Humphry Davy: Science and Power*, Blackwell, Oxford (1992).
[2] Alexander Marcet is presumed to be Jane Marcet's "friend" referred to in the Preface, see p. xxii, note 2.
[3] Dr. John Yelloy, a medical friend of her husband's, encouraged her with the advice that she should avoid "the familiarity which derogates from the Dignity of Science, and the abstruseness which has a tendency to make it forbidding..." and that "it is better to elevate

she explains in the Preface, her aim was to help beginners to understand the principles of the subject so that they might share her excitement and "see the wonders of nature in a new light".

Jane Marcet had in fact already written a science book aimed at young people: those who were beginning physics. She had used it for teaching her sister Sarah, but it had not yet been published[1]. Each topic was covered in a dialogue between a rigorous but kindly tutor, Mrs. B., and two young pupils: thirteen year-old Emily was clever and diligent, while Caroline, although also intelligent, was more extrovert in her enthusiasm but less disciplined in her thought. Jane Marcet planned to write her chemistry book along the same lines.

The contents of *Conversations on Chemistry* seem to have been based on the course which Alexander Marcet gave his wife, and on the illustrative experiments which they worked through together. But it was she who cast them in the form of lifelike dialogues, which reflect her own enthusiasm, intellectual rigour and personal charm. And Jane Marcet was not always the one who received the help in their scholarly projects; she also furthered her husband's research by editing his scientific papers (which his publishers considered to be written in too flowery a style) and then by proof-reading them.

The process of writing the book spanned two pregnancies, which went smoothly and resulted in the birth of Frank in May 1803 and of Frederick in April 1805. The book was finished that summer, accepted by Longman and Co. and (although this may amaze today's readers) published within three months. The price for the two bound volumes was 10 shillings. The first print run was 1000 copies, and no financial investment by the Marcets was required. On December 7, 1805 Alexander Marcet wrote that "when it was all done, we had a most large party".

However large and jubilant the party, it is unlikely that either the Marcets or their guests could have appreciated the importance of the occasion. The book was well reviewed[2] and was used with acclaim in girls' schools as well as by the women for whom the book was written. But not all her readers were female. Her most famous fan was Michael Faraday; he had come across the book in his teens, when he was a bookbinder's apprentice. He studied it after work, and repeated as many of the experiments as he had the money for. He later claimed that the book was

---

the minds of young ladies too high, than to depress them too low". See Crellin, J.K., *J.Chem.Educ.* **56** 59 (1979).

[1] See Preface, p. xxiv, note 2.
[2] See Appendix 1, p. 157, for a contemporary review.

the source of his enthusiasm for chemistry.[1] Jane Marcet was assiduous in trying to update each edition of her book; but despite her keeping in close touch with the work of Davy, and later of Faraday, at the Royal Institution, as well as taking part in discussions with her many scientific guests, her revisions were not always complete[2]. The last London edition, the "sixteenth", also containing her own revisions, was published in 1853 when she was eighty-four, bringing the total English sales[3] to 20,000. Comparison of the scientific and linguistic revisions which appear in successive volumes would offer rewarding sidelights on the changes which occurred during the period from 1806 to 1853, not only in chemistry and in social matters[4], but also in industrial science and in English usage.

The book was also very popular[5] in America, where it ran to twenty three editions. But some of these were published in the name of various American editors, who made appreciable alterations to the text, and sometimes included exercises for the student readers. Often no reference was made to its origin, and some title pages and book catalogues in both England and America implied that the fictitious tutor, thought to be a Mrs. Bryan[6], was herself the author. However, the book is claimed to have sold over 160,000 copies in America[7], and to have had as powerful an influence there as in Britain on the teaching of chemistry both to women and men It has been said that when a student asked Thomas Jefferson[8] how he should set about learning chemistry, the reply was "Read Mrs. Marcet's book"[9]. There were also two French editions.

Memory of the importance of *Conversations on Chemistry* survived into the first half of the twentieth century, particularly in the USA. The

---

[1] After Jane Marcet's death, her surviving children asked Auguste de la Rive (a relation by marriage and himself a scientist) to write an article about her work. He had heard of Faraday's youthful use of her book, and wrote to ask him if it this were true, and if so, whether he might quote it. Part of Faraday's reply is given in Appendix 2. Both letters are printed in full in *The Selected Correspondence of Michael Faraday*, ed. L. Pearce Williams, Cambridge University Press (1971). The reference to de la Rive's article is given by Armstrong, see note 14(ii).

[2] See Knight, D., *Ambix* **33** 94 (1986).

[3] I am most grateful to Dr. Michael Bott, of the University of Reading, UK for giving me the sales figures, and for telling me that there were in fact only fifteen editions, as the seventh never appeared.

[4] See note 1 above.

[5] Accounts of the American editions are given by e.g. (i) Smith, E.F. in *Old Chemistries*, McGraw-Hill, New York (1927) and (ii) Armstrong, E.V. in *J.Chem.Educ.*, 1938, p. 53.

[6] It has been said that Jane Marcet modelled Mrs. B. on her friend Mrs. Margaret Bryan, a headmistress who taught scientific subjects and wrote a book on astronomy, but the evidence for this does not seem to have been stated.

[7] Lindee, S., *Isis*, **81** 9 (1991).

[8] President of the United States of America 1801-1809.

[9] I am grateful to Tom Pasteur for this information.

University of Pennsylvania holds a copy of almost every American edition in its Edgar Fahs Smith Memorial Collection, and in 1938 a dramatic production was staged of one of the Conversations[1]. In Britain, however, Jane Marcet's name is familiar mainly to those who study the development of scientific education rather than to practising chemists or teachers. When the compiler of this selection was first interested in chemistry, and for many decades thereafter, she had never heard of Jane Marcet. She was, however, well acquainted with Mrs. B. and Caroline, snippets of whose interchanges were commonly used as chapter-head quotations in books written in the 1930's with such titles as *The Boys' Book of Chemical Experiments*. (The authors doubtless thought that Emily's more solemn contributions would have less appeal to their young readers.)

It would have been even more difficult for the gathering at the publication party to foresee that Jane Marcet's literary work would go far beyond chemistry. In 1816, she produced a companion volume on Political Economy, which was also highly acclaimed. This was followed by the publication of her earlier book on Natural Philosophy, by *Conversations* on topics which reflected her wide range of interests, and by stories for younger, and indeed for very young, children. Many of these had a sizeable content of elementary science. Until 1833, she did not use her name on the title page, but wrote instead "By the Author of Conversations on Chemistry". During her fifty-two years as an author, she published over thirty titles (not counting revised editions) and a game with cards and counters designed to teach English Grammar. Although these later works will also provide a fascinating, and surely most enjoyable, field for some future Marcet scholar, they must not be allowed to deflect our attention from focussing on her first published work.

The publication of *Conversations on Chemistry* was clearly a dramatic turning point in Jane Marcet's life; but did it also have any significance in a wider context? What types of people might want to read such a book, and how had they been educated? Was it usual, or even acceptable, for women of that period to have intellectual interests, especially scientific ones? What previous books had aimed to popularise chemistry, or indeed other sciences; and were any of these earlier books written in dialogue form? Can we, two centuries later, see why the book was so successful?

No answers to these questions can be discussed in isolation, since they depend on a number of interwoven historical threads, such as: the development of new scientific ideas, the spread of these ideas among non-specialists, and the inclusion of women in intellectual life and particularly

---

[1] Conversation XIX from the sixth English edition of 1819, *On the Muriatic and Oxymuriatic Acids; and on the Muriats*. See p.128-9 for excerpts from the first edition.

in the new public awareness of science. Any attempt to disentangle these areas of progress must start well before the eighteenth century, and will inevitably result in a very broad-brushed sketch; but a rough impression of the background to Jane Marcet's work is perhaps better than none at all.

As all human societies seem to have felt the need to construct frameworks of mythology, religion or science to help them to make sense of their surroundings, it is impossible to pin-point the start of "modern science". But great progress was made during the intense intellectual activity of the Renaissance, which often involved the "rebirth" of the appreciation of classical learning. In science, however, this period represented a rebirth of the use of reason, rather than a return to the knowledge which was available to the classical philosophers. Indeed, as the Middle Ages waned, some people started to question many of the views which had been accepted since classical times, including those of Aristotle about the natural world. Building on the work of astrologers and alchemists, the new natural philosophers extracted rational ideas, together with valuable technical know-how, from the mediaeval cauldron of science, mysticism, magic and quackery. As the centuries passed, new instruments were invented and new discoveries made. Thanks to the skills of the lens-grinders who had learned to make spectacles for the privileged, the power of the human eye was greatly extended, not only for reading and fine craftsmanship, but also for probing distances which had previously been too large or too small for our visual range. The telescope firmly established that the sun, and not the earth, was at the centre of our planetary system; and the microscope revealed the intricate but previously invisible structures of plants, of insects and of even smaller animals. The alchemists' successors no longer sought only to transmute base metals into gold or to discover the elixir of life but also performed experiments designed to elucidate the nature of matter and to find out what happens to it during such changes as combustion and dissolution. As they wished to tell their fellow investigators about the observations they had made, and the conclusions they had drawn from them, they reported their results in clear language which aimed to describe and explain their work, rather than to conceal it in a cloud of mystical symbolism, as many of the alchemists had done. But the replacement of the high-flown, and sometimes spiritual, aims of alchemical work by the more down-to-earth approach of the chemists took place very gradually. Alchemy survived well into the seventeenth century as was practised, alongside their better-known work, by such key thinkers as Isaac Newton, John Locke and Robert Boyle.

At first, these new-style investigators communicated with each other mainly by letter. But then they began to have meetings to discuss their findings, forming small groups which later grew into learned societies. In

London, the gathering which was to become the Royal Society first met in 1645; but the earliest such group, the Neapolitan *Academy of the Secrets of Nature* was formed eighty-five years earlier. By 1666, similar academies had also been founded in Rome, Florence and Paris.

The last four decades of the seventeenth century were dramatically fruitful for two Fellows of the newly founded Royal Society. Isaac Newton's work on mechanics, gravitational astronomy and optics transformed these subjects and put them on a quantitative basis which has served physics well for at least two centuries; even today modifications are needed only for systems in which the parameters are extreme, for example when dealing with interstellar distances or sub-atomic masses. Robert Boyle earned himself the sobriquet "Father of Chemistry" in a different way. He did not put his subject on a mathematical, or even factual, foundation but he did free it from some of its former shackles. In his *Sceptical Chymist* of 1661, Boyle[1] argued convincingly against the accepted view that all matter is made up of a small, fixed number of elements. The most popular number at that time was four (the famous Earth, Air, Fire and Water) which had been suggested by Empedocles and supported by Aristotle about two millennia previously. There were also some medically-minded alchemists who thought that all matter was composed of three principles (which they confusingly called salt, sulphur and mercury, although they distinguished them from the actual substances of the same names). *The Sceptical Chymist* contains the proposal that any substance should be regarded as elemental until it had been broken down into something simpler. This sensible suggestion, which originated with Helmont, was used by Boyle partly as an "Aunt Sally" and partly as a fuel for his contention that theoretical conclusions should be firmly based on experimental evidence. But, after more than a century, the idea was endorsed by Lavoisier in his definition of a chemical element; and it was rationalised, rather than overtaken, by John Dalton's atomic theory of 1804. However, since Dalton's atoms did not find immediate favour with Davy, they did not appear in the first edition of *Conversations on Chemistry*.

In 1667, a historic incident occurred at the Royal Society; its members voted (against considerable opposition), to accede to the unprecedented demand from a woman that she be invited to one of their meetings. The flamboyant Margaret Cavendish, Duchess of Newcastle, known to her many opponents and to posterity as "Mad Madge", was a vociferous exponent of education for women, and perhaps the first to emphasise that

---

[1] See *The Works of Robert Boyle,* ed. Hunter, M. and Davies, E. R., vol. 2, Pickering and Chatto, London (1999).

their education should include science[1]. But as her style of dress and manners were so brazenly unconventional, and her views so far ahead of her time, it is not surprising that she seems to have had little influence. A much more persuasive attempt to extend scientific knowledge to women was the publication in 1686 of a book[2] by the Frenchman Fontenelle, translated into English two years later as *The Plurality of Worlds*. Its aim was to initiate women into the delights of astronomy, whilst providing some entertainment, and possibly also instruction, for male readers. This book was probably the first scientific work to be written expressly for the non-specialist and it remains one of the most appealing. It was widely translated; and its many imitations would suggest that its influence persisted for much of the following century.

From the late seventeenth century onwards, there was continued progress, and maybe even acceleration, both of scientific activity and of its dissemination to people who were very different from the experimenters themselves. Leisured gentlemen were eager to keep abreast of the new ideas, and some educated ladies sought to learn enough to talk about scientific topics with their husbands and to instruct their children. Scientific writing flourished at a variety of levels[3].

As well as presenting their own ideas in learned books, scientists now wrote reports of the proceedings of the meetings at which they presented their work, and so founded the first scientific journals. They also published books which were aimed to be understood by a broader public. Their down-to-earth writing provoked the contempt of some literary figures of the day, such as Joseph Addison. Leaflets summarising astronomy and Newtonian physics were printed cheaply and formed a major topic of conversation at the coffee houses which were widely frequented by men of the intelligentsia. Science was also spread by textbooks for military cadets, handbooks for apprentice technologists, and more popularly by articles in magazines and periodicals and as entries in a variety of encyclopaedias and technical dictionaries. Some popularising scientific books, articles,

---

[1] An account of the influence of the new optical instruments on the scientific interests of women is given by Meyer, G.D., *The Scientific Lady in England, 1650 -1760*, University of California at Berkeley (1955).

[2] Fontenelle, B. deB. de, *Entretiens sur la Pluralité des Mondes* (1686), ed. Shackleton, R., Oxford University Press (1955), trans. Behn, A. *The Plurality of Worlds* (1688). Aphra Behn had no previous experience of either science or translation (but she had written plays, described as "coarse, lively and humorous", and had worked as a government spy for England against Holland).

[3] See Rousseau, G.D., *Science books and their readers in the eighteenth century*, in Rivers, I., ed. *Books and their Readers in Eighteenth-Century England*, Leicester University Press, (1982).

magazines[1] and even poems were written expressly for women; and problems in mathematical physics appeared regularly in the *Ladies' Diary*, a scientific and literary annual[2] which ran from 1704 to 1740. Such publications were not unique to England. In 1739, there appeared an anonymous translation[3] entitled *Sir Isaac Newton's Philosophy explained for the Use of Ladies*, written by Algarotti and published in Italian two years previously.

Science books were also written for children. Some of these were clear and matter-of-fact, such as the posthumously-published[4] *Elements of Natural Philosophy* which the philosopher John Locke (d.1704) wrote for "a young gentleman whose education he had very much at heart". Later, and much more fanciful, was Newbery's popular account[5] of the antics of the juvenile Lilliputian *Tom Telescope*, depicted in an illustration as prancing on the table while he lectured with immense authority to a mixed audience of varied age and high social class.

Later in the century, some astronomy books of intermediate level[6] were published for women who already knew the basics of the subject and had some mathematical expertise. So can we assume that such women were plentiful in England and accepted by society? It seems we cannot. But there were indeed a few very clever women, who had been educated at home in unusually enlightened families; and some of these did have scientific interests. There were probably many more who were skilled owners of small telescopes and microscopes. And an increasing number of

---

[1] *The Athenian Mercury*, ed. John Dunton, was published twice weekly from 1690 to 1697 for both men and women, while the *Free Thinker*, ed. Ambrose Philip, which ran from 1718 to 1721, also twice weekly, was for women only.
[2] Ed. John Tipper.
[3] See *Sir Isaac Newton's Philosophy explained for the Use of Ladies, in Six Dialogues on Light and Colour, from the Italian of Sig. Algarotti*, E. Cave, London (1739). The translator is now known to be the English polymath, Elizabeth Carter, who had been educated at home by her father.
[4] Published by Thomson and Dampier, London (1720).
[5] The publisher, John Newbery, is almost certain to have been the author, although in Oxford the book is catalogued under the authorship of "Telescope, T." Its full title is *The Newtonian System of Philosophy, adapted to the Capabilities of Young Gentlemen and Ladies, and familiarised and made entertaining by Objects with which they are initially acquainted: Being the Substance of Six Lectures read to the Lilliputian Society by Tom Telescope, A.M.* Dated 1761, it is described as a revised edition.
[6] For example Leadbeater, C., *Astronomy: or, The True System of the Planets Demonstrated*, Wilcox, and Heath, London (1727) and Charlton, J., *The Ladies' Astronomy and Chronology*, London (1735). Neither author makes any concession to lack of either basic knowledge or continuing commitment.

women wished to be aware of the latest scientific advances and attended with enthusiasm such lectures as were open to the public of either sex.[1]

As the eighteenth century proceeded, many of the conventions of society came under scrutiny and, following the ideas of Jean-Jacques Rousseau in France, there was a move towards a less rigid education, for both boys and girls. Some children were lucky enough to have lessons which included scientific observations and discussions. A number of groups of literary women[2] emerged, including the mocked "bluestockings"[3]; but there were other sets of female friends who wrote in modest obscurity[4]. Their writings show that at least some of them were aware of scientific activity[5] although there is no evidence that any of them had substantial scientific knowledge. However, the very existence of female literary groups may have eased the way for the progressive education which an increasing number of girls, including Jane Haldimand, were able to enjoy.

It may seem surprising that the some of the developments in British science went hand-in-hand with the modest improvement in educational opportunities for women; but both were furthered by the increasing number of educated men who vigorously challenged some traditional non-scientific institutions, such as the economy and the church. These reformers were not all leisured gentlemen. Some of the new leaders of the recent industrial revolution had often been both practical scientists and men of learning,[6] so that even at the end of the century any perceived gap between the humanities and the experimental sciences was like a hair-crack as compared with the situation today. And of the growing number of those

---

[1] As Meyer has pointed out (see p. ix, note 1), it is significant that some of the books for ladies were written by, and the lectures were given by, those who made and sold the optical instruments which were recommended.
[2] See (a) Myers, S.H., *The Bluestocking Circle*, Clarendon Press, Oxford (1990). (b) Small, H., *Times Lit. Supp.*, April 6, 2001.
[3] The Oxford English Dictionary explains that the word "bluestocking" was coined after a Jane Montague decided to spend some of her leisure in literary discussion with a few friends, both women and men, rather than frittering it away on games of cards. One guest habitually appeared wearing his grey (or "blue") worsted stockings rather than black silk ones. Although the term first seems to have referred merely to sartorial informality, it later implied that the women's literary interests were pretentious. The word is still used in both English and American, and applied to any female intellectual, regardless of her subject. Nowadays, it need not be derogatory, but when it is used in this way, it often means dowdy and prim-minded.
[4] See Reeves, M.E., *Pursuing the Muses: Female Education and Nonconformist Culture, 1700 -1900*, Leicester University Press (1997).
[5] I am grateful to Julia Saunders for telling me about Anna Aikin's poem *The Mouse's Petition*, a plea from a mouse awaiting one of Joseph Priestley's experiments on respiration. See *Poems* [A. L. Aikin], J. Johnson, London 5th edn, London (1777).
[6] See Uglow, J., *The Lunar Men*, Faber and Faber, London (2002).

who dissented from the tenets of the established church, many wished also to improve the educational system by introducing new subjects such as science and by offering more opportunities both to rich girls, and to the poor of both sexes. Some progressives, like the Unitarian minister Joseph Priestley, questioned not only current ecclesiastical and educational views, but also probed accepted scientific ideas by performing chemical experiments himself[1]. There was as yet no rift between science and religion; any advance in unravelling the workings of nature was believed to further man's understanding of the activities of the Creator. So it was not surprising that science should gradually to take its place among the subjects suitable for the formal education of boys and also for the few girls who attended the progressive schools which were beginning to be set up by the Unitarians.

In England, young men who were members of the established church could study the sciences at the Universities of Oxford and Cambridge. For the further scientific education of Dissenters, "Philosophical Institutes" were established, first at Warrington, and then at Manchester and Hackney. But neither the universities nor the institutes were open to women.

The new surge in scientific education produced a demand for instructional books as well as for popularising literature suitable for ladies and for other leisured self-improvers. At the beginning of the eighteenth century many of these books dealt with the work of Newton in astronomy, mechanics or optics, and they imparted or assumed some mathematical expertise. Others were purely descriptive accounts of nature; they were often restricted to plants and animals, but sometimes included geology, meteorology and physical geography. John Locke, followed by "Tom Telescope", covered all these areas, and treated human perception and understanding as branches of natural history.

Although the living world inspired books with beautiful illustrations, and descriptions of greater literary elegance than were usually found in books about physics, there was little organized system within the life sciences before the work of Carl von Linné (who used his Latinized name, Linnaeus) in the middle of the eighteenth century. The Linnaean classification of flowering plants by the number and structure of their reproductive organs, together with the introduction of a nomenclature which named both the genus and the species of each plant, produced a framework in which living things could be identified and docketed. Thanks to Linnaeus, the observational pastime of natural history was now poised to become an ordered science. Popular books on botanical classification,

---

[1] See (i) Golinski, J., *Science as Public Culture: Chemistry and Enlightenment in Britain, 1760 -1820*, Cambridge University Press (1992); (ii) Watts, R., *Gender, Power and the Unitarians in England, 1760 -1860*, Addison-Wesley Longman, Harlow, Essex (1998).

including Erasmus Darwin's highly florid 1155-line poem[1], followed; but many critics thought the sexual basis of the Linnaean system to be unsuitable for female readers, or indeed for any youthful ones.

So far, no attempt had been made to popularize chemistry, for the very good reason that the subject had not yet acquired an adequate systematic foundation. The chemists of the early and middle of the eighteenth century were, however, far from idle[2] and this period produced such distinguished names as: Bernouilli of Switzerland, Boerhaave of Holland, Cullen and Black of Scotland, Cavendish and Priestley of England and Scheele of Sweden. Many of these experimenters worked with gases, first learning how to collect them and transfer them from one vessel to another, and then probing the properties of these elusive substances, many of them colourless, odourless and tasteless, in the hopes of finding out what actually happens when something burns. But many other types of investigation were also carried out: studies of the behaviour of acids and the formation of salts, the extraction of organic acids from plant and animal material, and observational reaction chemistry of the type "let's see what happens when we distil this with that".

The second half of the eighteenth century in Europe saw a ferment of ideas, and it is not surprising that France, which was then the source of educational reform and political revolution, should be the spearhead also of chemical processes. Many of the new ideas arose from the work of Berthollet, Chaptal and Fourcroy; and they culminated in the elegant experiments, interpretative insights and lucid expositions of Lavoisier, a rich aristocrat who was privileged to have the money to commission expensive instruments for precise quantitative work and an energetic young wife who was also his amanuensis, graphic designer, translator and research assistant. Few of his contemporaries in other countries had similar advantages and some, such as Priestley, felt it proper to use apparatus so simple that his experiments could be repeated by the average educated householder.

The French system of chemistry contained two major innovations. One was binary nomenclature, reminiscent of the Linnaean classification of plants and animals; instead of referring to substances by traditional, individual names, Lavoisier devised a binary system, which indicated the chemical composition of the substance. For example, while the old name "Epsom Salts" suggested a salt-like substance once found at Epsom, the new term "Sulphat of Magnesia" conveyed to a chemist of the period that it could be made by adding a Sulphat group to Magnesia. More

---

[1] Darwin, E., *The Botanic Garden*, J. Johnson, London (1791).
[2] See Holmes, F. L., *Eighteenth Century Chemistry as an Investigative Enterprise*, University of California, Berkeley, (1989).

fundamentally, Lavoisier established that a component of air, which he called "oxygen", was essential for combustion. Several chemistry textbooks appeared in France at that time, and some of these appeared (very rapidly by today's standards) in English. By far the clearest, most elegant and most influential was Lavoisier's *Elements of Chemistry*[1]. He addressed his ideas to practicing chemists, but mainly to the younger ones, believing that only they were flexible enough to accept his innovations. Chemistry is indeed fortunate that this book was published before the French Revolution of 1789, which was to cost Lavoisier his life nearly five years later.

By the very end of the eighteenth century, the French chemists had made their subject systematic enough to be offered to non-specialist readers. The first of these more general books were intended for the use of medical students, technicians, school children and others who needed to learn basic chemistry. They dealt mainly with facts, rather than ideas; and even such authors as James Parkinson[2], who wrote persuasively of the rewards of studying chemistry, did not manage to convey the excitement of the subject. The forms used were textbook, compendium, encyclopaedia and "catechism" (a bald series of questions and answers designed for rote-learning using the same method as employed to teach religious dogma.). None of these books seems to have been written to provide pleasure or to stimulate the intellectual curiosity of readers such as literary gentlemen, women or inquisitive children. We do not know if Jane Haldimand read any of these books, but we do know that her particular interest in chemistry started only after hearing Davy's lectures at the Royal Institution, which he joined at the turn of the century, at about the time she got married. During the period when she was writing *Conversations on Chemistry*, her familiarity with the ideas of Lavoisier may well have been consolidated by reading his work.

We can well understand the need for a book to explain the new chemical principles to the appreciable number of women in Davy's audience, most of whom would have had no education in science. To many of them, the few available textbooks would have seemed too formal to be accessible, let alone enjoyable. Jane Marcet, having already written an elementary book on Newtonian physics for girls, had the experience to fill

---

[1] Lavoisier, A. *Traité Elementaire de la Chimie*, Paris (1789) ; Kerr, R., trans. *Elements of Chemistry*, Creech, Edinburgh (1790). Kerr complains that although he received the French edition only in mid September, the publishers wanted the book to appear for the start of the academic year, six weeks later.

[2] See Parkinson, J., *The Chemical Pocket Book*, Symonds, London, 2nd. edn. (1802). The author, like many chemists of his time, was also a physician. He wrote about the complaint which he called "shaking palsy", now known as Parkinson's disease.

this gap; and, despite her modesty in the preface, she may well have had the confidence to suppose that with her new enthusiasm and her husband's support she could complete a second book. It is not surprising that for *Conversations on Chemistry* she chose to use the same dialogue form as in her earlier *Conversations on Natural Philosophy*, and even the same three characters; but she made Emily and Caroline slightly older because she thought that no-one should embark on chemistry without first having acquired a basic knowledge of physics. But it is worth considering why she opted for this form in the first place.

During the period when Jane Haldimand was bringing up her younger siblings, she would surely have been aware of a number of educational books in conversational form. The genre was then quite fashionable, and had been sufficiently flexible, throughout its long tradition, to allow authors considerable freedom of both purpose and style[1]. She exploited these advantages with such skill that it now seems the obvious form for her to have used.

Educational dialogues certainly have a long and distinguished pedigree. The use of written conversations for communicating ideas goes back to Ancient Greece[2]. Plato (427 – 327 BC) who probably first used, and indeed immortalized, the dialogue form is thought to have admired the somewhat earlier work of Sophron of Syracuse, who wrote one-act sketches in rhythmic prose for dramatic performance. Written dialogues provide great scope for variety: of style and treatment, for level of readership, and for type of subject. Although they always have a strong central theme, they allow for asides, as might be made during a real conversation, in a lecture or, less gracefully, in modern editorial notes. Plato himself wrote for students of philosophy not only to publicize the ideas of his own teacher, Socrates[3], but also to illustrate how a student could be taught to think by skillful use of questions and answers in debate. In dialogues from later antiquity, fictional characters are often used and the subject of the discussion is not always so elevated; one long series of dialogues focuses on food.

The first scientific dialogues[4] appeared in the seventeenth century as disputations between persons of similar status, but of different views. Their aim was to persuade other scientists to discard currently accepted views in

---

[1] See Myers, G., *Fictions for Facts: The Form and Authority of the Scientific Dialogue*, Hist. Sci. 30 221 (1992).
[2] See Howatson, M., ed. *The Oxford Companion to Classical Literature*, Oxford University Press (1989).
[3] So far as we know, Socrates himself wrote nothing.
[4] I am extremely grateful to Julia Saunders of Wolfson College, Oxford for permission to use material from her unpublished article: *Arts and Science in Dialogue: Popular Science and Literary Forms in Eighteenth Century England*.

favour of ones more firmly based on observation and rational thought. In Galileo's *Dialogue Concerning the Two Chief World Systems*[1] (1632) two very learned characters, together with a less-gifted companion, debate whether the sun or the earth is the centre of the universe, in as even-handed a way as possible. (Galileo had hoped, albeit unsuccessfully, that this approach might hide his own strong preference for the sun from the church authorities). Nearly thirty years later, Boyle[2] chose to write his *Sceptical Chymist* as a similar dialogue between a progressive and a traditionalist, although he had no need to conceal his own views. The characters in these two books are used mainly to express opinions and to propel the argument; and Boyle's disputants, in particular, are lacking in personality. To the modern reader, both works would seem to be very hard going for anyone other than a devoted specialist.

Still later in the century came Fontenelle's *Entretiens sur la Pluralité des Mondes* (1686)[3]. Written for (predominantly female) non-specialist readers, the two characters are a natural philosopher with romantic views about both science and women, and his aristocratic young hostess. She would seem to be the ideal pupil, rich, elegant, attractive, modest and eager to learn, very intelligent and with a mind totally uncluttered by incorrect facts, traditional misconceptions or indeed by any scientific content. He flirts gently with this paragon whilst training her mind by Socratic dialogue, filling her brain with current ideas of cosmology and expanding her soul with his rapture at both the beauty of the universe and the joys of observing it. The philosopher acts as narrator, but the writing is so skillful (and so well translated) that the frequent use of such phrases as "said I" or "replied the Marchioness" never seems to jar. The characters are given full, plausible, and likeable personalities; and Fontenelle allows their conversations to branch out from the main theme of a discussion, without ever losing sight of it. It is easy to understand why this appealing work was so popular in its time, and so influential thereafter.

In the eighteenth century, a number of popularizations of science were written in dialogue form. The works[4] of Harris and of Algarotti (1737) also feature discussions of astronomy between a philosopher and a marchioness, and would seem to be direct imitations of Fontenelle, although in a far less elegant style; and the level of expertise expected from Harris's heroine makes for very unrelaxed reading. More recreational is Pluche's

---

[1] See Finocchiano, M. A., ed., trans. *Galileo on the World Systems*, University of California, Berkeley (1977).
[2] See p. viii, note 1.
[3] See p. ix, note 2.
[43] See (i) Harris, J., *Astronomical Dialogues Between a Gentleman and a Lady in a Pleasant, Easy and Familiar Way*, London (1719), and Algarotti, p. x, note 3.

Nature Displayed[1] in which a young knight learns natural history from conversations with a well-informed count, and a somewhat pompous prior. After the first discussion, the countess also appears, and is allowed to contribute to discussions which involve domestic activities such as gardening, cooking and rearing silkworms. Although the dialogues are presented in dramatic form, the characters seem much less three-dimensional than those of Fontenelle. As the century progressed, dialogues continued to be used in popular science, but with less aristocratic characters. Both James Ferguson[2] and Benjamin Martin[3] wrote in dramatic form, each depicting a young man who returns home after studying at university (in one case Cambridge, and in the other, probably Oxford) to find his sister eager to share some of his newly acquired knowledge. The characters are almost plausible, and their conversations include pleasantries and digressions, such as the expression of regret that girls have no chance of university education and of anxiety about whether scientifically educated girls would be socially acceptable. The "young gentlemen and ladies" to whom these books are addressed are presumably thoughtful teenagers. "Tom Telescope"[4] also appeared during this period, but it was written in narrative form, apparently for slightly younger readers, and the digressions often have a moral slant. The characters are in no way plausible; but the fact that the diminutive, all-knowledgeable Tom is no pasteboard character but a creation of fantasy may be one reason for the book's long-lived popularity[5].

As Jane Haldimand had herself been tutored in natural philosophy, she may have read "Tom Telescope"; and when she wrote her *Conversations on Natural Philosophy*, sometime in the two decades following her mother's death in 1782, she would probably have been aware of Maria Edgeworth's dialogue[6] between two gentlemen arguing for and against the education of women. She would also have been able to teach her young brothers to read using books containing simple conversations, usually between a mother and one or more of her children[7]. (One contains quite

---

[1] See Pluche, A.N., *Spectacle de la Nature* trans. Humphreys, S., *Nature Displayed, Being a Discourse on such Particulars of Natural History as were thought most proper to Excite the Curiosity and Form the Minds of Youth*, Pemberton et al., London, 2nd edn. (1733).
[2] See Ferguson, J., *The young gentleman and lady's astronomy*, Millar and Cadell, London (1768).
[3] See Martin, B., *The young gentleman's and lady's philosophy*, Owen, London 2nd edn. (1772).
[4] See p. x, note 5.
[5] The latest edition appeared in 1829.
[6] Edgeworth, M., *Letters for Literary Ladies*, J. Johnson, London (1795).
[7] For example (i) Fenn, E., *Cobwebs to catch flies*, John Marshall, London (1783) (ii) Aikin, A., and Barbauld, A. L. *Evenings at Home*, J. Johnson, London (1793).

lively dialogues using words of not more than three letters, thought to be suitable for readers aged three[1].) She would therefore have had ample opportunity to appreciate the great flexibility of the dialogue form for educational use; fictional conversations allow the author freedom to introduce opposing points of view (often in response to questioning), to confirm salient points by repetition, to explore side-alleys, to create distinct and plausible characters who help to keep the reader's interest, and even to insert moral exhortations. And with her experience in the schoolroom, she would have appreciated that a skillful author can use a dialogue to teach good habits of thought.

So by the time that Jane Marcet came to write *Conversations on Natural Philosophy*, the use of fictional conversations to further the scientific education of women (and others) was far from new. And the fact that she was one of the first to write lively chemistry for the general reader depended on her being in the right places at the right time: near enough to London to get to lectures which were open to women; in a household with a library, a lab and a husband with chemical interests and contacts; with children who needed lessons which included science; and active during the years following Lavoisier's new systematization of chemistry. As a young woman, she was lucky to have robust health, and the money to provide enough leisure to make use of her fortunate situation. She made no claims to be an expert in the subject or to have made any discoveries herself; but as a recent learner she wanted to share her knowledge, so that her readers might share her enthusiasm. But why was the book of this modest non-specialist so successful during her lifetime? And indeed why should we wish to celebrate it two centuries later? Is there some special ingredient, or has she put her own mark on the dialogue as an educational device?

A full attempt to answer such questions would need to be illustrated by examples from Jane Marcet's writing; and to quote any of her work verbatim in this introduction would be like disclosing the end of a thriller whilst recommending it to a friend. Much of the fun of reading it would be spoiled. But familiarity with guide-book descriptions can enhance a visit to an unknown place; and enjoyment of a book can be increased if the reader, while looking for particular features, finds yet others which have not been mentioned.

While the early popularizing dialogues are set in the elegant gardens of French chateaux, Jane Marcet's take place in a schoolroom, a most suitable venue for realistic chemistry lessons. Her main aims seem to be realism and clarity, and her down-to-earth attitude confirms her conviction that chemistry is above all a practical subject. If students are unable to do

---

[1] See p. xvii, note 7(i).

experiments themselves or even to watch demonstrations, then at least they should get convincing experiences second-hand from a chemistry book which is as realistic as it is possible to make it. Her work contains extremely well-printed engravings of her own exquisite drawings of apparatus, sometimes including hands to show how pieces are to held and manipulated; and she makes Mrs. B. explain clearly what each part was for, and how it is used. The reader is made to feel a member of the group, able to enjoy the colour changes, to exclaim with delight or disgust at any flashes, bangs and stinks, and to learn such experimental skills as handling acids and blowing glass. As an experienced teacher, Jane Marcet makes it clear that the practical demonstrations have been prepared before the lesson, so as to avoid time-waste, disappointment or frustration. She uses practical work, especially in Caroline's somewhat slap-happy hands, to promote some awareness of health and safety, although perhaps less rigorously than would be required by today's legislation, or indeed than would be desired by the other occupants of the house.

Although Jane Marcet was convinced that practical work plays a key role in chemical learning, the subject matter of her *Conversations* was much wider. She made it clear that chemistry was not a collection of established facts, but a changing set of ideas postulated to fit the ever-growing body of experimental evidence. So a substance then thought to be an element might soon be broken down, and thus shown to be a compound; or it might, as happened with 'caloric', later be found to have a quite different status. And if explanations which had been proposed for observations seemed improbable, they should be treated as suspect until more progress had been made. There was much chemistry which was not yet known, and there was much which the tutor chose not to teach, either because she thought it too confusing for her pupils, or because, as she frankly admitted, her own knowledge was not as complete as she would have wished. Jane Marcet handled innovations in chemistry with the same robust realism. Although she much admired Lavoisier's new system of naming chemical substances, she warned that traditionalists might resist such gross changes, and advocated sensitivity and restraint. (Her sensible views on this topic should surely be required reading for all chemists concerned with advising on changes in chemical nomenclature). Always a pragmatist, she saw no problem in the use of Réaumur and Fahrenheit scales of temperature in different countries; after all, it was simple to learn how to convert a value from one scale to the other. Jane Marcet's attractive lack of dogmatism gives her readers the sense of a rapidly developing subject which is absent from earlier books of popularization of science.

Jane Marcet certainly presented chemistry "in the round", but were her characters equally plausible? She herself admits in the Preface that some of the remarks made by the pupils are unrealistically acute for their supposed age, but that this "fault" was necessary to avoid the "tedious repetitions" which would have been necessary in real life. Naturally, she crafted Mrs. B. into her ideal of a tutor for beginners, making her kindly but rigorous, and disciplined about both systematic teaching and clarity of thought. In any educational dialogue, it is useful to have an able and virtuous pupil like Emily who could produce sensible answers, ask pertinent questions, and give an account of recently assimilated material in her own words. Although these two characters are indeed conventional, they are no mere pasteboard figures; an adult reader might well feel uncomfortably ambivalent about Emily, who would be a joy to teach, but infuriating as a tutorial partner. Some writers of dialogue might have been tempted to add a second pupil who gets everything wrong in order to provide ample opportunity for correction (just as, after more than one and a half centuries, the ineptitudes of the fictitious Freddy Jones were used to hone the critical skills of pupils studying physics under the Nuffield scheme). Jane Marcet's Caroline, however, is no such fool, and she follows no conventions, literary or otherwise. She is seldom used to ask idiot questions, but rather to keep the discussions lively and to show the importance of intellectual discipline. Sometimes she even allowed to take the mickey out of Mrs. B., though with the utmost courtesy. So Emily's competence is never allowed to bore the reader for long. Such is Jane Marcet's skill that the reader may have trouble in remembering that the text is fictional, rather than a verbatim transcript of conversations which took place between real people. Although there are attempts at characterization in several earlier dialogues of popular science, not even Fontenelle can approach Jane Marcet in the creation of such plausible personalities.

Jane Marcet also applied her realism to make her *Conversations* ring true, not as formal discussions, nor as everyday chatter, but as credible tutorials. Mrs. B. is even concerned that her pupils do not get tired or overloaded with information. She makes it clear that each lesson has been planned to cover certain material in a particular order, but within this framework, Jane Marcet makes much use of one of the great strengths of the conversational form; the digression. In this way, the discussion is made lighter and livelier, the reader's interest is kept awake, and fertile shoots can grow out of the central chemical theme. Some of these digressions are purported to come from Mrs.B.'s initiative, while others arise in response to questions from her pupils. Their subjects are widely divergent, but all are educational and all are related to chemistry. There is advice about methods of study, and there are many instances of guidance in clear thinking. Connections are made between chemistry and other topics, particularly those with which the pupils are familiar in daily life, and in their own personal interests. As well as stimulating the pupils with accounts of new developments in chemistry, Mrs. B. frequently tells them of the benefits which the subject had recently brought to society through changes to industry and agriculture. The subsequent lively discussions of social and economic issues gave them a better understanding of the lifestyle of the people who worked in these areas. One or two lessons contain guidance on the sensitive issue of how a young woman can be well-informed and enthusiastic about chemistry, without appearing to be a pedant or a know-all.

It would seem that Jane Marcet was no primary innovator. Although she was the first person to attempt a popularization of what was largely Lavoisier's system of chemistry, there had been earlier popularizations of other sciences; and her chosen form of the dialogue, with its characterization and digression, was hardly new. So how can we account for the popularity and lasting influence of her *Conversations on Chemistry*? A modern reader might see her achievement as two-fold. She had a precise and innovative vision of the numerous strands which she felt necessary for a chemical education; and she was able to convey this vision in her writing with such clarity, conviction and realism that the reader feels an onlooker at, indeed almost a participant in, a superb tutorial. But perhaps she also achieved something else, more difficult and, some readers might think, even more important. Two centuries later, much of the book still makes an extremely enjoyable read.

JANE MARCET'S PREFACE[1]

In venturing to offer to the public, and more particularly to the female sex, an Introduction to Chemistry, the author, herself a woman, conceives that some explanation may be required; and she feels it the more necessary to apologize for the present under-taking, as her knowledge of the subject is but recent, and as she can have no real claims to the title of chemist.

On attending, for the first time, experimental lectures, the author found it almost impossible to derive any clear or satisfactory information from the rapid demonstrations which are usually, and perhaps necessarily, crowded into popular courses of this kind. But frequent opportunities having afterwards occurred of conversing with a friend[2] on the subject of chemistry, and of repeating a variety of experiments, she became better acquainted with the principles of that science, and began to feel highly interested in its pursuit. It was then that she perceived, in attending the excellent lectures delivered at the Royal Institution, by the present Professor of Chemistry[3], the great advantage which her previous knowledge of the subject, slight as it was, gave her over others who had not enjoyed the same means of private instruction. Every fact or experiment attracted her attention, and served to explain some theory to which she was not a total stranger; and she had the gratification to find that the numerous and elegant illustrations, for which that school is so much distinguished, seldom failed to produce on her mind the effect for which they were intended.

Hence it was natural to infer, that familiar conversation was, in studies of this kind, a most useful auxiliary source of information; and more

---

[1] MI, v-x. The location of each excerpt in the first (1806) edition of Jane Marcet's book is shown by the volume number (indicated by M1 or M2) followed by the span of pages (usually in Arabic numerals, but for front matter in lower case Roman.)

[2] The "friend" was almost certainly her physician husband, Alexander Marcet, who had his own chemical laboratory and was a keen experimentalist.

[3] The lectures of the handsome and charismatic Professor (later Sir Humphry) Davy are said to have drawn large, appreciative audiences, and contemporary pictures show that women were well-represented.

especially to the female sex, whose education is seldom calculated to prepare their minds for abstract ideas, or scientific language[1].

As, however, there are but few women who have access to this mode of instruction; and as the author was not acquainted with any book that could prove a substitute for it, she thought it might be useful for beginners, as well as satisfactory to herself, to trace the steps by which she had acquired her little stock of chemical knowledge, and to record, in the form of a dialogue, those ideas which she had first derived from conversation.

But to do this with sufficient methods, and to fix upon a mode of arrangement, was an object of some difficulty. After much hesitation, and a degree of embarrassment, which, probably, the most competent chemical writers have often felt in common with the most superficial, a mode of division was adopted, which, though the most natural, does not always admit of being strictly pursued— it is that of treating first of the simplest bodies, and then gradually rising to the most intricate compounds.

It is not the author's intention to enter into a minute vindication of this plan. But, whatever may be its advantages or inconveniences, the method adopted in this work is such, that the young pupil, who should occasionally recur to it, with a view to procure information on particular subjects, might often find it obscure or unintelligible; for its various parts are so connected with each other as to form an uninterrupted chain of facts and reasonings, which will appear sufficiently clear and consistent only to those who may have the patience to go through the whole work. Or have previously devoted some attention to the subject.

It will, no doubt, be observed, that in the course of these conversations, remarks are often introduced, which appear much too acute for the young pupils, by whom they are supposed to be made. Of this fault the author is fully aware. But, in order to avoid it, it would have been necessary to omit a number of useful illustrations, or to submit to such minute explanations and frequent repetitions, as would have rendered the work much less suited to its purpose.

---

[1] Middle-class girls in late 18th century England were educated with a view to an early, but suitable, marriage. It was desirable for them to have a little knowledge of literature, history, geography, religion and maybe botany together with some accomplishment in music, art and French, particularly if accompanied by modesty. But following Rousseau in France, English reformers such as Erasmus Darwin and the Edgeworth family were beginning to advocate radical changes, including even co-education. There was little desire that girls, like their brothers, be enslaved by classical languages, but rather that the education of both sexes should be based more on the child's own experience, with less dependence on books; and it should include the sciences. Wives would then be able to have well-informed discussions with liberally educated husbands and to take a responsible interest in the education of their children; and the rare single women, and even some of the married ones, could develop serious intellectual interests of their own.

In writing these pages, the author was more than once checked in her progress by the apprehension that such an attempt might be considered by some, either as unsuited to the ordinary pursuits of her sex, or ill justified by her own recent and imperfect knowledge of the subject. But, on the one hand, she felt encouraged by the establishment of those public institutions, open to both sexes, for the dissemination of philosophical knowledge, which clearly prove that the general opinion no longer excludes women from an acquaintance with the elements of science ; and, on the other, she flattered herself that whilst the impressions made upon her mind, by the wonders of Nature studied in this new point of view, were still fresh and strong, she might perhaps succeed the better in communicating to others the sentiments she herself experienced.

It will be observed, that, from the beginning of the work, it is taken for granted that the reader has previously acquired some slight knowledge of natural philosophy[1], a circumstance, indeed, which appears very desirable. The author's original intention was to commence this work by a small tract, explaining, on a plan analogous to this, the most essential rudiments of that science. This idea she has since abandoned, but the manuscript[2] was ready, and might, perhaps, have been printed at some future period, had not an elementary work of a similar description[3], under

---

[1] Following Aristotle, the term *natural philosophy* originally implied the study of all natural phenomena amongst which Locke in 1704 included even gods and angels. But during the eighteenth century its meaning became limited; first, biology was separated off, and then chemistry (which at that time claimed heat as an elementary substance). The term physics, used as early as 1715, coexisted with natural philosophy for over a century. But although Mary Somerville discussed "branches of physics" in 1835, much as we would today, the term "natural philosophy" was in common use until the mid-nineteenth century; and even in 1888, Davy was described as a "natural philosopher".

[2] *"Conversations on Natural Philosophy in which the Elements of that Science are familiarly explained and applied to the comprehension on young people; by the author of Conversations on Chemistry, and on Political Economy"*, was the first book which Jane Marcet wrote although it was not published until 1819. The characters are the same as those in "Conversations on Chemistry", which the work is intended to precede, but the style is slightly more formal. Jane Marcet may have felt that the younger pupils needed a firmer rein; or perhaps she herself relaxed as she became more accustomed to writing. Possibly, too, as an avowed non-mathematician, she was more comfortable with the "less quantitative" discipline of chemistry. The book covers Mechanics, Astronomy, Hydrostatics and Optics, but not heat, which Mrs. Marcet treats as a chemical element, "caloric" (see pp. 6, 14.)

[3] *Scientific dialogues: intended for the instruction and entertainment of young people in which the first principles of natural and experimental philosophy are fully explained (1805) [anon].* In the first edition two somewhat pasteboard pupils, Charles and James, converse with a similarly characterless tutor. By 1812, the author's name was given as the Rev. Jeremiah Joyce and the tutor had been replaced by the father of the two pupils, now Charles and Emma. Slight gender and personality traits are discernable, in that Charles is interested in definitions of technical terms, the interpretation of diagrams, and in numerical problems

the title of "Scientific Dialogues," been lately pointed out to her, which, on a rapid perusal, she thought very ingenious, and well calculated to answer its intended object.

---

while Emma asks about broader concepts, such as "But can philosophy be comprehended by children as young as we are? I thought it had been the business of men and of old men too." The book is aimed at readers aged ten or eleven and in his preface dated 1800 the author acknowledges his debt to "Mr. Edgeworth's" *Practical Education*, which was in fact written mainly by Richard Edgeworth and his daughter Maria. In addition to the subjects covered by Jane Marcet, the seven small volumes include pneumatics, magnetism, and electricity (but not heat).

Selections from

VOL. I

ON SIMPLE BODIES

CONVERSATION I

ON THE GENERAL PRINCIPLES OF CHEMISTRY

The scope of chemistry[1]

MRS. B.

HAVING[2] now acquired some elementary notions of NATURAL PHILOSOPHY[3], I am going to propose to you another branch of science, to which I am particularly anxious that you should devote a share of your attention. This is CHEMISTRY, which is so closely connected with Natural Philosophy, that the study of the one must be incomplete without the knowledge of the other; for it is obvious that we can derive but a very imperfect idea of bodies from the study of the general laws by which they are governed, if we remain totally ignorant of their intimate nature.

CAROLINE

To confess the truth, Mrs. B., I am not disposed to form a very favourable idea of chemistry, nor do I expect to derive much entertainment from it. I prefer those sciences that exhibit nature on a grand scale, to those which are confined to the minutiae of petty details. Can the studies which we have lately pursued, the general properties of matter, or the revolutions of the heavenly bodies, be compared to the mixing up of a few insignificant drugs?

MRS. B.

I rather imagine that your want of taste for chemistry proceeds from the very limited idea you entertain of its object. You confine the chemist's

---

[1] M1, 1-7. The numbers following M1 and M2 here and elsewhere refer to the page numbers in the first edition, of volumes 1 and 2 respectively.
[2] By the 16th edition, "Having...." had been replaced by "As you have..." (Some readers of this note may have been chastised for similar grammatical errors.)
[3] See p. xxiv, notes 1 and 2.

1

laboratory to the narrow precincts of the apothecary's shop, whilst it is subservient to an immense variety of other useful purposes. Besides, my dear, chemistry is by no means confined to works of art.[1] Nature also has her laboratory, which is the universe, and there she is incessantly employed in chemical operations. You are surprised, Caroline; but I assure you that the most wonderful and the most interesting phenomena of nature are almost all of them produced by chemical powers. Without entering therefore into the minute details of practical chemistry, a woman may obtain such a knowledge of the science, as will not only throw an interest on the common occurrences of life, but will enlarge the sphere of her ideas, and render the contemplation of nature a source of delightful instruction.

CAROLINE
If this is the case, I have certainly been much mistaken in the notion I had formed of chemistry. I own that I thought it was chiefly confined to the knowledge and preparation of medicines.

MRS. B.
That is only a branch of Chemistry, which is called Pharmacy; and though the study of it is certainly of great importance to the world at large, it properly belongs to professional men, and is therefore the last that I should advise you to study.

EMILY
But did not the chemists formerly employ themselves in search of the philosopher's stone, or the secret of making gold?

MRS. B.
These were a particular set of misguided philosophers, who dignified themselves with the name of Alchemists, to distinguish their pursuits from those of the common chemists, whose studies were confined to the knowledge of medicines. But, since that period, chemistry has undergone so complete a revolution, that, from an obscure and mysterious art, it is now become a regular and beautiful science, to which art is entirely subservient. It is true, however, that we are indebted to the alchemists for many very useful discoveries, which sprung from their fruitless attempts to make gold, and which undoubtedly have proved of infinitely greater advantage to mankind than all their chemistrical pursuits.

The modern chemists, far from directing their ambition to the

---

[1] Jane Marcet used the word *art* to mean an acquired human skill, often (as here) contrasted with natural phenomena.

imitation of one of the least useful productions of inanimate nature, aim at copying almost all her operations, and sometimes even form combinations, the model of which is not to be found in her own productions. They have little reason to regret their inability to make gold which is often but a false representation of riches, whilst by their innumerable inventions and discoveries, they have so greatly stimulated industry and facilitated labour, as prodigiously to increase the luxuries as well as the necessaries of life.[1]

EMILY

But I do not understand by what means chemistry can facilitate labour; is not that rather the province of the mechanic?

MRS. B.

There are many ways by which labour may be rendered more easy, independently of mechanics; but even the machine the most wonderful in its effects, the steam engine, cannot be understood without the assistance of chemistry.[2] In agriculture, a chemical knowledge of the nature of soils, and of vegetation, is highly useful; and in those arts which relate to the comforts and conveniences of life, it would be endless to enumerate the advantages which result from the study of this science.

CAROLINE

But, pray, tell us more precisely in what manner the discoveries of chemists have proved so beneficial to society?

MRS. B.

That would be an unfair anticipation; for you would not comprehend the nature of such discoveries and useful applications, so well as you will do hereafter. Without a due regard to method, we can not expect to make any progress in chemistry. I wish to direct your observation chiefly to the chemical operations of Nature; but those of Art are certainly of too high importance to pass unnoticed. We shall therefore allow them also some share of our attention.

---

[1] The frequent references to the benefits of chemical industry to human well-being suggest that the author's *"Conversations on Political Economy"* (1816) was based on long-standing socioeconomic interests.

[2] The *Advertisement* to the 16th (1853) edition states: "In the Tenth edition of the work a Conversation was added on the Steam-engine, a machine which contributes so abundantly to the wealth, power, and happiness of this country, that the Author considers it deserving of particular attention". In *The Seasons: Autumn* (1854), a series of scientifically-based stories "for very young children", she included an account of the generation and use of steam-power in the 35 pages she devoted to the steam-boat.

EMILY

Well, then, let us now set to work regularly. I am very anxious to begin.

MRS. B.

The object of chemistry is to obtain a knowledge of the intimate nature of bodies, and of their mutual action on each other. You find therefore, Caroline, that this is no narrow or confined science, which comprehends every thing material within our sphere.

CAROLINE

On the contrary, it must be inexhaustible; and I am at a loss to conceive how any proficiency can be made in a science whose objects are so numerous.

MRS. B.

If every individual substance was formed of different materials, the study of chemistry would indeed be endless; but you must observe that the various bodies in nature are composed of certain elementary principles, which are not very numerous.

CAROLINE

Yes; I know that all bodies are composed of fire, air, earth, and water[1]; I learnt that many years ago.

MRS. B.

But you must now endeavour to forget it. I have already informed you what a great change chemistry has undergone since it has become a regular science. Within these thirty years especially, it has experienced an entire revolution, and it is now proved that neither fire, air, earth, nor water, can

---

[1] The Four Elements hypothesis, attributed to the Greek philosopher Empedocles (c. 495-435 BC), was formalized and publicized by Aristotle (384-322 BC). Although from the Middle Ages onwards there was discussion about the number and nature of the few "principles" supposed to constitute matter; it remained generally agreed that there was a small, fixed number of such elements. It was only in 1661 that a disputant in Robert Boyle's *"Sceptical Chymist"* proposed for the sake of argument that any substance should be regarded as an element if it had so far eluded attempts to separate it into simpler components, a suggestion which eventually held sway. Despite this most rational idea, children were apparently brought up to accept the two-thousand-year-old Four Elements hypothesis for at least another 160 years. (We ourselves might wonder if, two millennia from now, the young will still be taught that there are exactly seven colours in the rainbow).

be called elementary bodies.[1] For an elementary body is one that cannot be decomposed, that is to say, separated into other substance; and fire, air, earth, and water, are all of them susceptible of decomposition.

EMILY

I thought that decomposing a body was dividing it into its minutest parts. And if so, I do not understand why an elementary substance is not capable of being decomposed, as well as any other.

MRS. B.

You have misconceived the idea of *decomposition*; it is very different from mere *division*: the latter simply reduces a body into parts, but the former separates it into the various ingredients, or materials, of which it is composed.

Simple or elementary bodies[2]

MRS. B.

The elementary substances of which a body is composed, are called the *constituent* parts of that body; in decomposing it, therefore, we separate its constituent parts. If, on the contrary, we divide a body by chopping it to pieces, or even by grinding or pounding it to the finest powder, each of these small particles will still consist of a portion of the several constituent parts of the whole body: these we call the *integrant* parts; do you understand the difference?

EMILY

Yes, I think, perfectly. We *decompose* a body into its *constituent* parts; and *divide* it into its *integrant* parts.

MRS. B.

Exactly so. If therefore a body consist of only one kind of substance, though we may divide it into its integrant parts, it is not possible to decompose it. Such bodies are therefore called *simple* or *elementary*, as

---

[1] Jane Marcet was greatly influenced by Lavoisier's *Elementary Treatise on Chemistry* (1789) which drew on laboratory and linguistic work of various chemists in the previous decades to establish chemistry as a rational, experimental and even quantitative science, served by a systematic nomenclature. The chemical elements, which now included oxygen, were defined as recommended by Boyle.

[2] M1, 8-10.

they are the elements of which all other bodies are composed. *Compound bodies* are such as consist of more than one of these elementary principles.

CAROLINE

But do not fire, air, earth, and water, consist, each of them, but of one kind of substance?

MRS. B.

No, my dear; they are every one of them susceptible of being separated into various simple bodies. Instead of four, chemists now reckon upwards of forty elementary substances. These we shall first examine separately, and afterwards consider in their combinations with each other. Their names are as follows:

Light, Caloric, Oxygen, Nitrogen, Hydrogen, Sulphur, Phosphorus, Carbone.

2 Alkalies Potash, Soda

10 Earths Lime, Magnesia, Strontites, Barytes, Silex, Alumine, Yttria, Glucina[1], Zirconia, Agustina[2]

25 Metals Gold, Platina, Silver, Mercury, Copper, Iron, Tin, Lead, Nickel, Zinc, Bismuth, Antimony, Arsenic, Cobalt, Manganese, Tungsten, Molybdenum, Uranium, Tellurium, Titanium, Chrome, Osmium, Iridium, Palladium, Rhodium.

CAROLINE

This is, indeed, a formidable list!

MRS. B.

Not so much as you imagine; many of the names you are already acquainted with, and the others will soon become familiar to you.

---

[1] Beryllium oxide, BeO. A modern reader, alert to the toxicity of beryllium compounds, might recoil from its original name, meaning sweet-tasting.
[2] No reference can be found to this substance, which, from its name, was presumably tasteless.

Chemical attraction[1]

MRS. B.

Before we proceed further, it will be necessary to give you some idea of chemical attraction, a power on which the whole science depends.

Chemical Attraction, or the *Attraction of Composition,* consists in the peculiar tendency which bodies of a different nature have to unite with each other. It is by this force that all the compositions, and decompositions, are effected.

EMILY

What is the difference between chemical attraction, and the attraction of cohesion, or of aggregation, which you often mentioned to us in former conversations?

MRS. B.

The attraction of cohesion exists only between particles of the *same* nature, whether simple or compound; thus it unites the particles of a piece of metal which is a simple substance, and likewise the particles of a loaf of bread which is a compound. The attraction of composition, on the contrary, unites and maintains in a state of combination particles of a *dissimilar nature*.........

EMILY

The attraction of cohesion, then, is the power which unites the integrant particles of a body; the attraction of composition that which combines the constituent particles. Is it not so?

MRS. B.

Precisely: and observe that the attraction of cohesion unites particles of a similar nature, without changing their original properties; the result of such an union, therefore, is a body of the same kind as the particles of which it is formed; whilst the attraction of composition, by combining particles of a dissimilar nature, produces new bodies, quite different from any of their constituent particles. If, for instance, I pour on the piece of copper, contained in this glass, some of this liquid (which is called nitric acid) for which it has a strong attraction, every particle of the copper will

---

[1] M1, 10-13.

combine with a particle of acid[1], and together they will form a new body, totally different from either the copper or the acid.

Do you observe the internal commotion that already begins to take place? It is produced by the combination of these two substances; and yet the acid has in this case to overcome not only the resistance which the strong cohesion of the particles of copper oppose to its combination with them, but also the weight of the copper, which makes it sink to the bottom of the glass, and prevents the acid from having such free access to it as it would if the metal were suspended in the liquid.

EMILY

The acid seems, however, to overcome both these obstacles without difficulty, and appears to be very rapidly dissolving the copper.

MRS. B.

By this means it reduces the copper into more minute parts, than could possibly be done by any mechanical power. But as the acid can act only of the surface of the metal, it will be some time before the union of these two bodies will be completed.

You may, however, already see how totally different this compound is from either of its ingredients. It is neither colourless like the acid, nor hard, heavy, and yellow, like the copper. If you tasted it, you would no longer perceive the sourness of the acid. It has at present the appearance of a blue liquid; but when the union is completed, and the water with which the acid is diluted is evaporated, it will assume the form of regular crystals, of a fine blue colour, and perfectly transparent. Of these I can show you a specimen, as I have prepared some for that purpose.

CAROLINE

How very beautiful they are, in colour, form, and transparency! Nothing can be more striking than this example of chemical attraction.

---

[1] It would be another 78 years before this reaction could be rationalised in terms of Arrhenius's theory of ionic dissociation, although the concept of an ion ("something moving") was suggested in 1834 by Faraday in order to account for electrolysis.

Attraction or affinity?[1]

MRS. B.
The term *attraction* has been lately introduced in to chemistry as a substitute for the word *affinity*, to which some chemists have objected, because it originated in the vague notion that chemical combinations depended upon a certain resemblance, or relationship, between particles that are disposed to unite; and this idea is not only imperfect, but *erroneous*, as it is generally particles of the most dissimilar nature, that have the greatest tendency to combine.

CAROLINE
Besides, there seems to be no advantage in using a variety of terms to express the same meaning; on the contrary it creates confusion; and as we are well acquainted with the term attraction in natural philosophy, we had better adopt it in chemistry likewise.

MRS. B.
If you have a clear idea of the meaning, I shall leave you at liberty to express it in the terms you prefer. For myself, I confess that I think the word attraction best suited to the general law that unites the integrant particles of bodies; and affinity better adapted to that which combines the constituent particles, as it may convey an idea of the preference which some bodies have for others, which the term *attraction of composition* does not so well express.

EMILY
So I think; for though that preference may not result from any relationship, or similitude, between the particles (as you say was once supposed), yet, as it really exists, it ought to be expressed.

MRS. B.
Well, let it be agreed that you may use the terms *affinity, chemical attraction, and attraction of composition*, indifferently, provided you recollect that they have all the same meaning.

---

[1] M1, 13.

Decomposition[1]

EMILY
I do not conceive how bodies can be decomposed by chemical attraction. That this power should be the means of composing them, is very obvious; but how it can at the same time produce exactly the contrary effect, appears to me very singular.

MRS. B.
To decompose a body, is, you know, to separate its constituent parts[2], which, as we have just observed, can never be done by mechanical means.

EMILY
No; because mechanical means separate only the integrant particles; they act merely against the attraction of cohesion.

MRS. B.
The decomposition of a body, therefore, can only be performed by chemical powers. If you present to a body composed only of two principles, a third, which has a greater affinity for one of them than the two first have for each other, it will be decomposed, that is, its two principles will be separated by means of the third body. Let us call two ingredients, of which a body is composed, A and B. If we present to it another ingredient C, which has a greater affinity for B than that which unites A and B, it necessarily follows that B will quit A to combine with C. The new ingredient, therefore, has effected a decomposition of the original body A B; A has been left alone, and a new compound, B C, has been formed.

EMILY
We might, I think, use the comparison of two friends, who were very happy in each other's society, till a third disunited them by the preference which one of them gave to the new-comer.

MRS. B.
Very well. I shall now show you how this takes place in chemistry.

---

[1] M1, 14-17.
[2] In a molecular solid, the *constituent parts* correspond to atoms, and the *integrant particles* to conglomerations of molecules.

Let us suppose that we wish to decompose the compound we have just formed by the combination of the two ingredients, copper and nitric acid: we may do this by presenting to it a piece of iron, for which the acid has a stronger attraction than for copper; the acid will consequently quit the copper to combine with the iron, and the copper will be what the chemists call *precipitated,* that is to say, it will return to its separate slate, and reappear in its simple form.

In order to produce this effect, I shall dip the blade of this knife into the fluid, and, when I take it out, you will observe that, instead of being wetted with a blueish liquid like that contained in the glass, it will be covered with a very thin pellicle of copper.

CAROLINE

So it is, really! But then is it not the copper, instead of the acid, that has combined with the iron blade?

MRS. B.

No; you are deceived by appearances: it is the acid which combines with the iron, and in so doing deposits the copper on the surface of the blade.

EMILY

But cannot three or more substances combine together, without any of them being precipitated?

MRS. B.

That is sometimes the case; but, in general, the stronger affinity destroys the weaker; and it seldom happens that the attraction of several substances for each other is so equally balances as to produce such complicated compounds.

It is now time to conclude our conversation for this morning. But, before we part, I must recommend you to fix in your memory the names of the simple bodies, against our next interview.

CONVERSATION II

ON HEAT AND LIGHT

The separation of heat from light[1]

CAROLINE
We have learned by heart the names of all the simple bodies, which you have enumerated, and we are now ready to enter on the examination of each of the successively. You will begin, I suppose, with LIGHT?

MRS. B.
That will not detain us long; the nature of light, independent of heat, is so imperfectly known, that we have little more than conjectnures respecting it.

EMILY
But is it possible to separate light from heat; I thought they were only different degrees of the same thing?

MRS. B.
They are certainly very intimately connected; yet it appears that they are distinct substances, as they can, under certain circumstances, be in a great measure separated. The most striking instance of this was pointed out by Dr. Herschell[2].

This philosopher discovered that heat was less refrangible than light; for in separating the different coloured rays of light by a prism (as we did some time ago), he found that the greatest heat was beyond the spectrum,

---

[1] M1, 18-22.
[2] Friedrich Wilhelm Herschel (1738 -1822) who in England became William, and later, Sir William. Using dark glasses to observe the sun safely, he found in 1800 that heat and light were differently refracted, the maximal heat being outside the red end of the visible spectrum, in the *infra-red* region. See, Cornell, E.S., *Annals of Science*, **3** 119-137 (1938).

at a little distance from the red rays, which you may recollect are the least refrangible.

EMILY

I should like to try that experiment.

MRS. B.

It is by no means an easy one: the heat of a ray of light, refracted by a prism, is so small, that it requires a very delicate thermometer to distinguish the difference of the degree of heat within and without the spectrum. For in this experiment, the heat is not totally separated from the light, each coloured ray containing a certain portion of it, though the greatest part is not sufficiently refracted to fall within the spectrum.

CAROLINE

Perhaps the different degrees of heat which the seven rays possess, may in some unknown manner occasion their variety of colour. I have heard that melted metals change colour according to the different degrees of heat to which they are exposed; might not the colour of the spectrum be produced by a cause of the same kind? Do let us try if we cannot ascertain this, Mrs. B.– ? I should like extremely to make some discovery in chemistry.

MRS. B.

Had we not better learn first what is already known? Surely you cannot seriously imagine that, before you have acquired a single clear idea of chemistry, you can have any chance of discovering secrets that have eluded the penetration of those who have spent their whole lives in the study of that science.

CAROLINE

Not much, to be sure, in the regular course of events; but a lucky chance sometimes happens. Did not a child lead the way to the discovery[1] of telescopes?

MRS. B.

There are certainly a few instances of this kind. But, believe me, it is infinitely wiser to follow up a pursuit regularly, than to trust to chance for your success.

---

[1] The telescope is said to have been discovered after someone, possibly a boy apprentice, who looked through two simple lenses which chanced to be of appropriate curvature and distance apart to magnify distant objects.

EMILY

But to return to our subject. Though I no longer doubt that light and heat can be separated, Dr. Herschel's experiment does not appear to me to afford sufficient proof that they are essentially different; for light, which you call a simple body, may likewise be divided into the various coloured rays; is it not therefore possible that heat may be only a modification of light?

MRS. B.

That is a supposition which, in the present state of natural philosophy, can neither be positively affirmed nor denied: it is generally thought that light and heat are connected with each other as cause and effect; but which is the cause, and which the effect, it is extremely difficult to determine. But it would be useless to detain you any longer on this intricate subject. Let us now pass on to that of HEAT, with which we are much better acquainted.

CAROLINE

Heat is not, I believe, amongst the number of simple bodies?

MRS. B.

Yes, it is; but under another name— that of CALORIC; which is nothing more than the principle, or matter of heat.— We suppose caloric to be very subtle fluid, originally derived from the sun, and composed of very minute particles, constantly in agitation and moving in a similar manner to light, as long as they meet with no obstacle.

But when these rays come in contact to the earth, and the various bodies belonging to it, part of them are reflected from their surfaces according to certain laws, and part enters into them.

CAROLINE

These rays of heat, or caloric, proceeding from the same source, and following the same direction, as the rays of light, bear a very strong resemblance to them.

MRS. B.

So much so that it often requires great attention not to confound them.

EMILY

I think there is no danger of that, if we recollect one great distinction— light is visible, and caloric is not.

MRS. B.
Very right. *Light* affects the sense of *Sight*; *Caloric* that of *Feeling*: the one produces *Vision*, and the other the peculiar sensation of *Heat*.

Caloric and heat[1]

MRS. B.
Caloric[2] is found to exist in a variety of forms, and susceptible to certain modifications, all of which may be comprehended under the four following heads:
1. FREE CALORIC,
2. SPECIFIC HEAT,
3. LATENT HEAT,
4. CHEMICAL HEAT.

The first, or FREE CALORIC, is also called HEAT OF TEMPERATURE; it comprehends all heat which is perceptible to the senses, and affects the thermometer.

EMILY
You mean such as the heat of the sun, of fire, of candles, of stoves; in short, of every thing that burns?

MRS. B.
And likewise of things that do not burn, as, for instance, the warmth of the body; in a word, all heat that is *sensible,* whatever may be its degree, or the source from which it is derived.

CAROLINE
What then are the other modifications of caloric? It must be a strange kind of heat that cannot be perceived by our senses?

MRS. B.
None of the modifications of caloric should properly be called *heat*; for heat, strictly speaking, is the sensation produced by caloric, on animated bodies, and this word should therefore be confined to express the

---

[1] M1, 22-27.
[2] This topic would nowadays be considered as physics rather than chemistry; but these few excerpts are included to illustrate the use of the concept of caloric in Jane Marcet's explanations.

sensation. But custom has adapted it likewise to inanimate matter, and we say *the heat of an oven, the heat of the sun,* without any reference to the sensation which they are capable of exciting.

It was in order to avoid the confusion which arose through thus confounding the cause and effect, that modern chemists adopted the new word *Caloric*, to express the principle which produces heat; but they do not yet limit the word *heat* (as they should do) to the expression of the sensation, since they still retain the habit of connecting this word with the three other modifications of caloric.

CAROLINE

But you have not yet explained to us what these other modifications of caloric are.

MRS. B.

Because you are not yet acquainted with the properties of free caloric, and you know that we have agreed to proceed with regularity.

One of the most remarkable properties of free caloric is its power of *dilating* bodies. This fluid is so extremely subtle, that it enters and pervades all bodies whatever, forces itself between their particles, and not only separates them, but, by its repulsive power, drives them asunder, frequently to a considerable distance from each other. It is thus that caloric expands or dilates a body so as to make it occupy a greater space than it did before.

EMILY

The effect of caloric on bodies therefore, is directly contrary to that of the attraction of cohesion; the one draws the particles together, the other drives them asunder.

MRS. B.

Precisely. There is a kind of continual warfare between the attraction of aggregation and the repulsive power of caloric; and from the action of these two opposite forces, result all the various forms of matter, or degrees of consistence, from the solid, to the liquid and aeriform state. And accordingly we find that most bodies are capable of passing from one of these forms to the other, merely in consequence of their receiving different quantities of caloric.

CAROLINE

That is very curious; but I think I understand the reason of it. If a great quantity of caloric is added to a solid body, it introduces itself between the particles in such a manner as to overcome, in a considerable degree, the attraction of cohesion; and the body, from a solid, is then converted into a fluid.

MRS. B.

This is the case whenever a body is melted; but if you add caloric to a liquid, can you tell me what is the consequence?

CAROLINE

The caloric forces itself in greater abundance between the particles of the fluid, and drives them to such a distance from each other, that their attraction of aggregation is wholly destroyed; and the liquid is then transformed into vapour.

MRS. B.

Very well; and this is precisely the case with boiling water, when it is converted into steam or vapour.

But each of these various states, solid, liquid, and aeriform admit of many degrees of density, or consistence, still arising (partly, at least) from the different quantities of caloric the bodies contain. Solids are of varying degrees of density, from that of gold, to that of a thin jelly. Liquids, from the consistence of melted glue, or melted metals, to that of ether, which is the lightest of all liquids. The different elastic fluids (with which you are not yet acquainted) admit of no less variety in their degrees of density.

EMILY

But does not every individual body also admit of different degrees of consistence, without changing its state?

MRS. B.

Undoubtedly; and this I can immediately show you by a very simple experiment. This piece of iron now exactly fits the frame, or ring, made to receive it; but if heated red hot, it will no longer do so, for its dimensions will be so much increased by the caloric that has penetrated into it, that it will be much too large for the frame.

The iron is now red hot; by applying it to the frame, we shall see how much it is dilated[1].

---

[1] It is perhaps surprising that Jane Marcet did not take this opportunity to mention the fitting of hot iron rims on to wooden cart wheels.

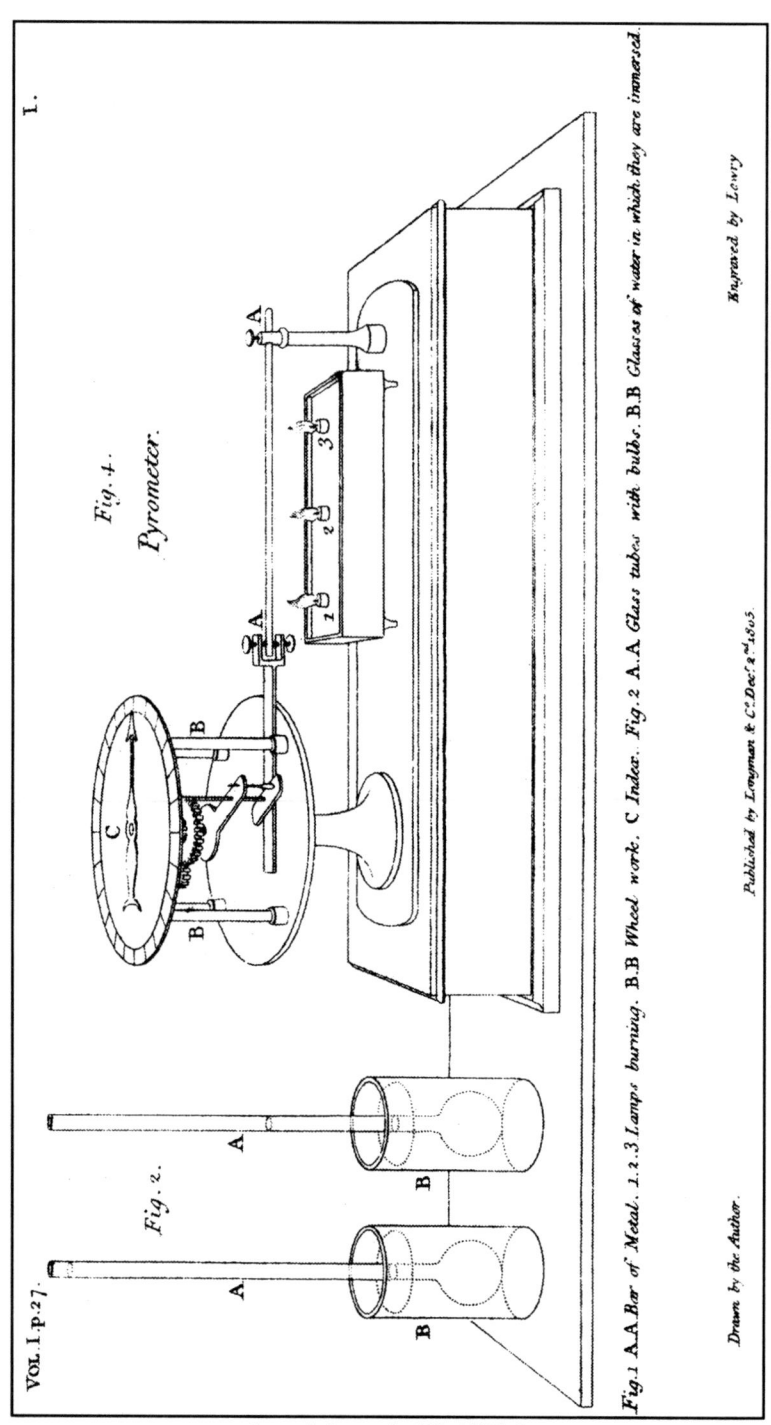

Jane Marcet's PLATE I

EMILY
Considerably so indeed! I knew that heat had this effect on bodies, but I did not imagine that it could be made so conspicuous.

MRS. B.
By means of this instrument (called a Pyrometer[1]) we may examine, in the most direct manner, the various dilations of any solid body by heat. The body we are now going to submit to trial is this small iron bar; I fix it to this apparatus (PLATE I, Fig.1.) and then heat it by lighting the three lamps beneath it: when the bar dilates, it increases in length as well as thickness; and, as one end communicates with this wheel-work, while the other end is fixed and immovable, no sooner does it begin to dilate than it presses against the wheel-work, and sets in motion the index, which points out the degrees of dilation on the dial-plate.

Thermometers and temperature scales[2]

MRS. B.
A thermometer therefore consists of a tube with a bulb, such as you see here, containing a fluid whose degrees of dilation and contraction are indicated by a scale to which the tube is fixed. The degree which indicates the boiling point, simply means that when the fluid is sufficiently dilated to rise to this point, the heat is such, that water exposed to the same temperature will boil. When, on the other hand, the fluid is so condensed as to sink to the freezing point, we know that water will freeze at that temperature. The extreme points of the scales are not the same on all thermometers, nor are the degrees always divided in the same manner. In different countries philosophers have chosen to adopt different scales

---

[1] The Oxford English Dictionary gives this use of the word *pyrometer* as dating from 1749 and Herschel used it in this sense in 1830. But in 1839, it was used, as it is today, to denote the ceramic cones and balls which potters used to assess the temperature of a kiln.
[2] M1, 30-32.

and divisions. The two thermometers most used are those of Fahrenheit[1], and of Reaumur; the first is generally preferred by the English, the latter by the French.

EMILY

The variety of scale must be very inconvenient, and I should think liable to occasion confusion, when French and English experiments are compared.

MRS. B.

This inconvenience is but very trifling, because the different graduations of the scales do not affect the principle on which thermometers are constructed. When we know, for instance, that Fahrenheit's scale is divided into 212 degrees, in which 32° corresponds with the freezing point, and 212° with the boiling point of water; and that Reaumur's[2] is divided into only 80 degrees in which 0° denotes the freezing point, and 80° that of boiling water, it is easy to compare the two scales together, and reduce the one into the other. But, for greater convenience, thermometers are sometimes constructed with both these scales, one on either side of the tube; so that the correspondence of the different degrees of the two scales, is thus easily seen. Here is one of these scales, (PLATE II. Fig. 3.) by which you can at once perceive that each degree of Reaumur's corresponds to 2¼ of Fahrenheit's division.

---

[1] D.G.Fahrenheit (1686-1736) was born of German parents in Danzig, and worked with an instrument maker in Amsterdam. He worked on producing a universal thermometric scale in the period 1708-17.

[2] The Frenchman R-A.F. de Réaumur (1683-1757) devised his scale for metal and porcelain technology.

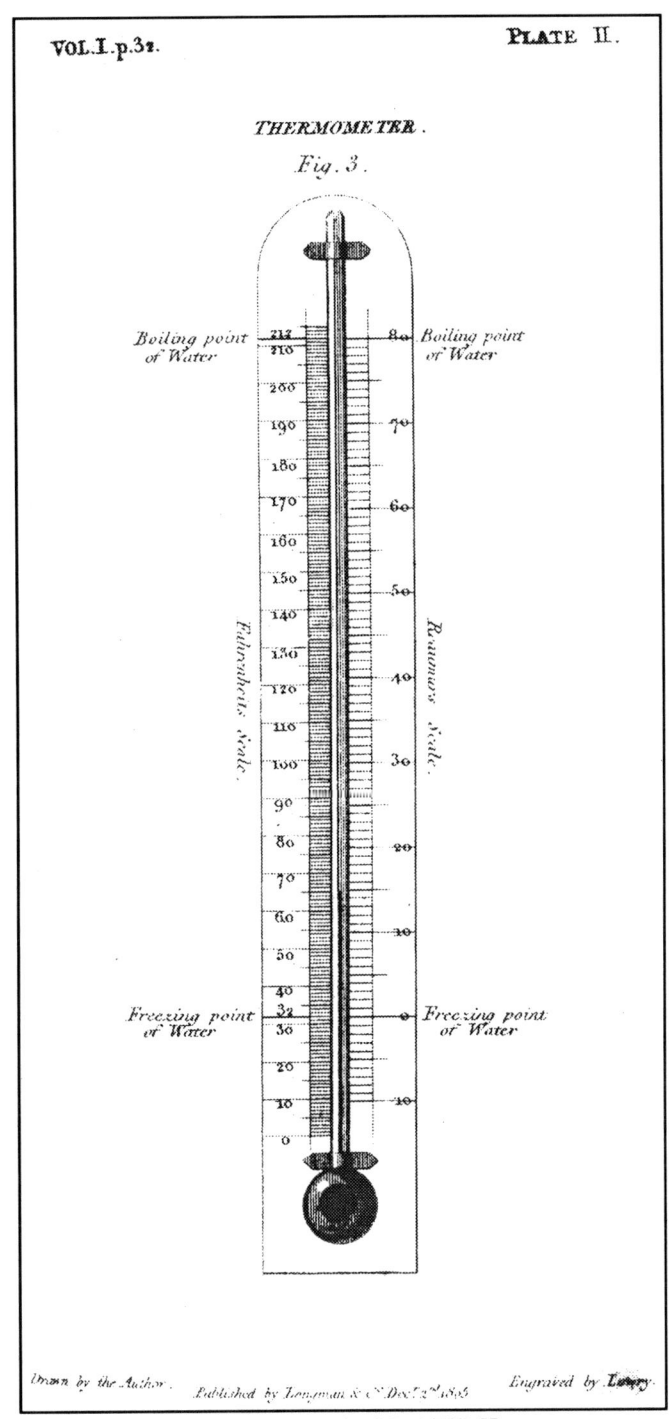

Jane Marcet's PLATE II

CONVERSATION III

CONTINUATION OF THE SUBJECT

'Hot' and 'cold' objects[1]

MRS. B.
...the constant tendency of free caloric to restore and equilibrium of temperature,..... when well understood, affords the explanation of a great variety of facts which appeared formerly unaccountable. You must observe, in the first place, that the effect of this tendency is gradually to bring all bodies that are in contact, to the same temperature. Thus, the fire which burns in the grate, communicates its heat from one object to another, till every part of the room has an equal proportion of it.

EMILY
And yet this [*leather*] book is not so cold as the [*marble*] table on which it lies, though both are at an equal distance from the fire, and actually in contact with each other, so that, according to your theory, they should be exactly of the same temperature?

CAROLINE
And the [*wooden*] hearth, which is much nearer the fire than the carpet, is certainly the colder of the two.

MRS. B.
If you ascertain the temperature of these several bodies with a thermometer (which is a much more accurate test than your feeling), you will find that it is exactly the same.

CAROLINE
But if they are of the same temperature, why should one feel colder than the other?

---

[1] M1, 45-47.

MRS. B.

The hearth and the table feel colder than the carpet or the book, because the latter are not such *good conductors of heat* as the former. Caloric finds a more easy passage through marble and wood, than through leather and worsted; the two former will therefore absorb heat more rapidly from your hand, and consequently give it a stronger sensation of cold than the two latter, although they are all of them really at the same temperature.

CAROLINE

So, then, the sensation I feel on touching a cold body, is in proportion to the rapidity with which my hand yields its heat to that body?

MRS. B.

Precisely; and, if you lay your hand successively on every object in the room, you will discover which are good, and which are bad conductors of heat, by the different degrees of cold you feel. But, in order to ascertain this point, it is necessary that the several substances should be of the same temperature, which will not be the case with those that are very near the fire, or those that are exposed to a current of cold air from a window or door.

CONVERSATION IV

ON SPECIFIC HEAT, LATENT HEAT, AND CHEMICAL HEAT

Steam heating[1]

EMILY
Pray do let us see the effect of latent heat returning to its natural state.

MRS. B.
For the purpose of showing this, we need simply conduct the vapour through this tube, into this vessel of sold water, where it will part with its latent heat and return to its liquid state—

EMILY
How rapidly the steam heats the water!

MRS. B.
This is because it does not merely impart its free caloric to the water, but likewise its latent heat. There is a large dye-house at Leeds, in which a great number of kettles are kept boiling by means of a single one which is heated by fire; the steam of this last, instead of being allowed to escape and be wasted, is conveyed through pipes into each of the other coppers, which are thus heated without any additional fuel.

CAROLINE
That is an admirable contrivance, and I wonder that it is not in common use.

MRS. B.
The steam kitchens, which are getting into such general use, are upon the same principle. The steam is conveyed through a pipe in a similar manner, into several vessels which contain the provisions to be dressed,

---

[1] M1, 96-98.

where it communicates to them its latent caloric, and returns to the state of water. Count Rumford[1] makes great use of this principle in many of his fire-places: his grand maxim is to avoid all unnecessary waste of caloric, for which purpose he confines the heat in such a manner, that not a particle of it shall unnecessarily escape; and while he economises the free caloric, he takes care also to turn the latent heat to advantage. It is thus that he is able to produce a degree of heat superior to that which is obtained in common fireplaces, though he employs but half the quantity of fuel.

EMILY

When the advantages of such contrivances are so clear and plain, I cannot understand why they are not universally used.

MRS. B.

A long time is always required before innovations, however useful, can be reconciled with the prejudices of the vulgar.

EMILY

What a pity it is that there should be a prejudice against new inventions; how much more rapidly the world would improve, if such discoveries were immediately, and universally, adopted!

MRS. B.

I believe, my dear, that there are as many novelties attempted to be introduced, the adoption of which would be prejudicial to society, as there are those that would be beneficial to it. The well-informed, though by no means exempt from error, have an unquestionable advantage over the illiterate, in judging what is likely or not to be serviceable; and therefore we find the former more ready to adopt such discoveries as promise to be really advantageous, than the latter, who, having no other test of the value of a novelty but time and experience, at first oppose its introduction. The well-informed are, however, frequently disappointed in their most sanguine expectations, and the prejudices of the vulgar, though they often retard the progress of knowledge, yet sometimes, it must be admitted, prevent the propagation of error.— But we are deviating from our subject.

---

[1] Benjamin Thompson (1753-1816).

Is caloric a substance?[1]

CAROLINE
Calorie appears to me a most wonderful element; but I cannot reconcile myself to the idea of its being a substance; for it seems to be constantly acting in opposition, both to the attraction of aggregation and to the laws of gravity; and yet you decidedly class it among the simple bodies.

MRS. B.
You are not at all singular in the doubts you entertain, my dear, on this point; for although caloric is now believed to be a real substance, yet there are certainly some strong circumstances which seem to militate against this doctrine.

CAROLINE
But do *you*, Mrs. B., believe it to be a substance?

MRS. B.
Yes, I do; but I am inclined to think, that its levity is, in all probability, only relative, like that of vapour, which ascends through the heavier medium, air.

CAROLINE
In that be the case, it would not ascend in a vacuum.

MRS. B.
In an absolute vacuum, perhaps, it would not. But as the most complete vacuum we can obtain is never perfect, we may always imagine the existence of some unknown invisible fluid, which, however light and subtle, may be heavier than caloric, and will gravitate in it. The fact has not, I believe, been yet determined by very decisive experiments; but it appears from some made by Professor Pictet[2], mentioned in his 'Essay on

---

[1] M1, 114-116.
[2] M-A. Pictet (1752-1825) gave a series of lectures in his native Geneva, Switzerland in 1790, supporting the views of Lavoisier. Alexander Marcet records (notebooks of 1801 and 1805, Marcet Collection, Archive Guy de Pourtalès, Etoy, Switzerland) that Pictet encouraged Jane Marcet to publish Conversations on Chemistry when he had heard her reading from it after he had dined with the Marcets.

Fire'[1], that heat has a tendency to ascend in the most complete vacuum which we are able to obtain.

EMILY.

But if there exists such a subtle fluid as you imagine, do you not think that chemists would have discovered it by some of its properties?

MRS. B.

It has been conjectured that light might be such a fluid; but I confess that I do not think it probable: for, as it appears by Dr. Herschell's experiment[2] that heat is less refrangible than light, I should be rather inclined to think it the heavier of the two. But, while you have so many well ascertained facts to learn, I shall not perplex you with conjectures. We have dwelt on the subject of caloric much longer than I intended, and I fear you will find it difficult to remember so long a lesson[3].

---

[1] Published in French in 1790, with an English translation (London, 1791) by W. B[elcombe].
[2] See p.12, note 2.
[3] Conversations II, III and IV take up nearly one hundred of the 325 pages of Volume I.

CONVERSATION V

ON OXYGEN AND NITROGEN

Combustion of wood and coal[1]

EMILY
And by what means can the two gasses, which compose the atmospheric air, be separated?

MRS. B.
There are many ways of analysing the atmosphere; the two gasses may be separated first by combustion.

EMILY
How is it possible that combustion should separate them?

MRS. B.
I must first tell you, that all bodies, excepting the earths and alkalies, have so strong an affinity for oxygen, that they will, in certain circumstances, attract and absorb it from the atmosphere; and in this case the nitrogen gas remains alone, and we thus obtain it in its simple gaseous state.

CAROLINE
I do not understand how a gas can be absorbed?

MRS. B.
The gas is not absorbed, but decomposed; and it is oxygen only, that is to say, the basis of the gas, which is absorbed.

CAROLINE
What then becomes of the caloric of the oxygen gas, when it is deprived of its basis?

---

[1] M1, 121-129.

MRS. B.

We shall make this piece of dry wood absorb oxygen from the atmosphere, and you will see what becomes of the caloric.

CAROLINE

You are joking, Mrs. B.– ; you do not mean to decompose the atmosphere with a piece of stick?

MRS. B.

Not the whole body of the atmosphere, certainly; but if we can make this stick absorb any quantity of oxygen from it, will not a proportional quantity of atmospherical air be decomposed?

CAROLINE

Undoubtedly: but if the wood has so strong an affinity for oxygen, as to attract it from the caloric with which it is combined in the atmosphere, why does it not decompose the atmosphere spontaneously?

MRS. B.

Because the attraction of aggregation of the particles of the wood, is an obstacle to their combination with the oxygen; for you know that the oxygen must penetrate the wood, in order to combine with its particles, and forcibly separate them in direct opposition to the attraction of aggregation.

EMILY

Just as caloric penetrates bodies?

MRS. B.

Yes; but caloric being a much more subtle fluid than oxygen, can penetrate substances much more easily.

CAROLINE

But if the attraction of cohesion between the particles of a body, counteracts its affinity for oxygen, I do not see how that body can decompose the atmosphere?

MRS. B.

That is now the difficulty which we have to remove with regard to the piece of wood. – Can you think of no method of diminishing the attraction of cohesion?

CAROLINE

Heating the wood, I should think, might answer the purpose; for the caloric would separate the particles and make room for the oxygen.

MRS. B.

Well, we shall try your method: hold the stick close to the fire – closer still, that it may imbibe the caloric plentifully; otherwise the attraction of cohesion between its particles may not be sufficiently overcome –

CAROLINE

It has actually taken fire, and yet I did not let it touch the coals; but I held it so very close, that I suppose it caught fire merely from the intensity of the heat.

MRS. B.

Or you might say, in other words, that the heat so far overcame the attraction of cohesion of the wood, that it was enabled to absorb oxygen very rapidly from the atmosphere.

EMILY

Does the wood absorb oxygen while it is burning?

MRS. B.

Yes; and the heat and light are produced by the caloric of the oxygen gas, which being set at liberty by the oxygen uniting with the wood, appears in its sensible form.

CAROLINE

You astonish me! Is it possible that the heat of a burning body should be produced by the atmosphere and not by the body itself?

MRS. B.

It is not precisely ascertained whether any portion of the caloric is furnished by the combustible body; but there is no doubt that by far the most considerable part of it is disengaged from the oxygen gas, when its basis combines with the combustible body.

CAROLINE

Since I have learnt this wonderful theory of combustion, I cannot take my eyes from the fire; and I can scarcely conceive that heat and light, which I always supposed to proceed from the coals, are really produced by the atmosphere and that the coals are only the instruments by which the decomposition of the oxygen gas is affected.

Prerequisites for Combustion[1]

EMILY

When you blow the fire, you increase the combustion, I suppose, by supplying the coals with a greater quantity of oxygen gas?

MRS. B.

Certainly; but of course no blowing will produce combustion, unless the temperature of the coals be first raised. A single spark, however, is sometimes sufficient to produce that effect; for, as I said before, when once combustion has commenced, the caloric disengaged is sufficient to elevate the temperature of the rest of the body, provided that there be a free access of oxygen. There are, therefore, three things required in order to produce combustion; a combustible body, oxygen, and a temperature at which the one will combine with the other[2].

---

[1] M1, 129.
[2] An early, and admirably clear, statement of what is now often known as the *fire triangle*.

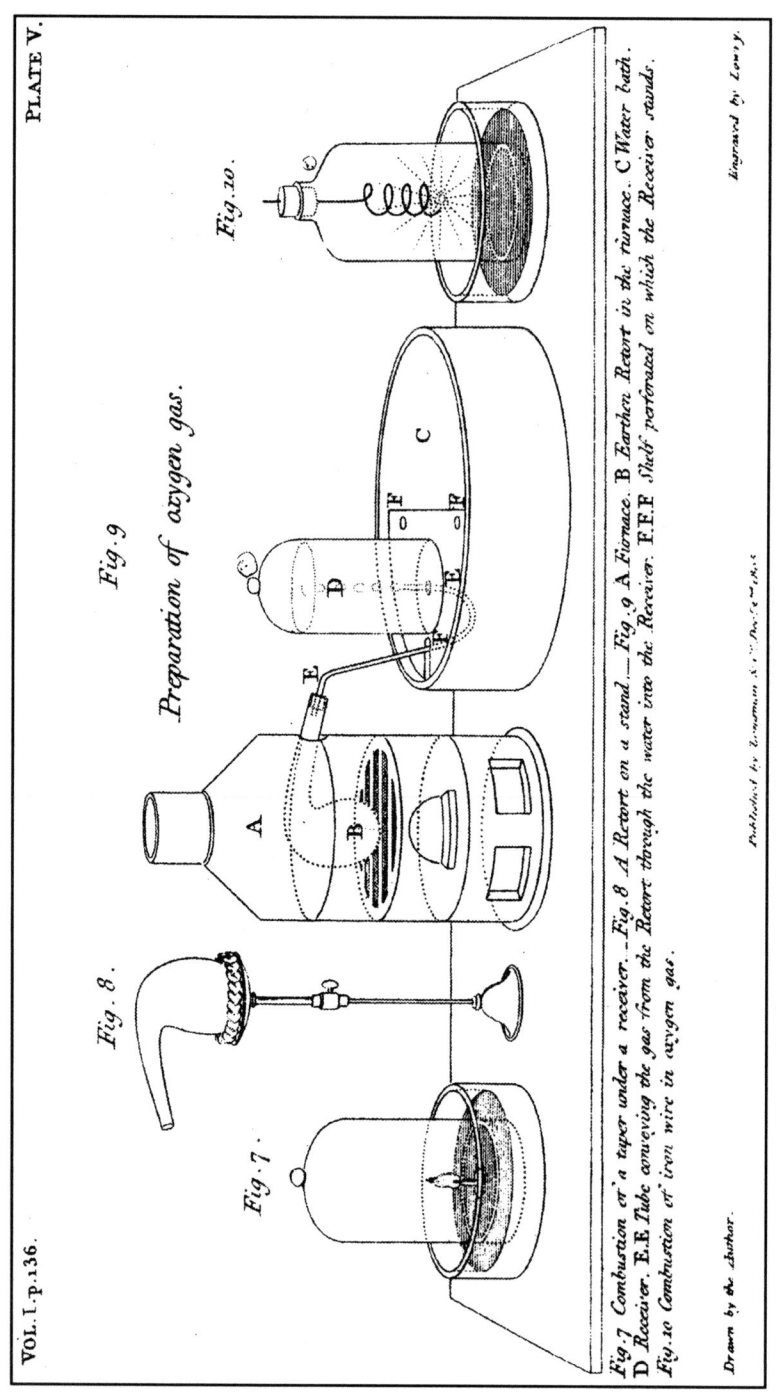

Jane Marcet's PLATE V

The separation of nitrogen from air[1]

EMILY
You said that combustion was one method of decomposing the atmosphere[2], and obtaining nitrogen gas in its simple state; but how do you secure this gas, and prevent it from mixing with the rest of the atmosphere?

MRS. B.
It is necessary for this purpose to burn the body in a closed vessel, which is easily done.- We shall introduce a small lighted taper (PLATE V. Fig.7) under this glass receiver, which stands in a bason of water, to prevent all communication with the external air.

CAROLINE
How dim the light burns already!– It is now extinguished.

MRS. B.
Can you tell us why it is extinguished?

CAROLINE
Let me consider- The receiver was full of atmospherical air; the taper, in burning within it, must have absorbed the oxygen contained in that air, and the caloric which was disengaged produced the light of the taper. But when the whole of the oxygen was absorbed, the whole of the caloric was disengaged; consequently the taper ceased to burn, and the flame was extinguished.

MRS. B.
Your explanation is perfectly correct.

EMILY
The two constituents of oxygen gas being thus disposed of, what remains under the receiver must be pure nitrogen?

---

[1] M1, 129-131.
[2] Readers may enjoy the modern drama *Oxygen* (Djerassi, C., and Hoffmann, R., Wiley-VCH, Weinheim, 2001) which is based on the rival claims for its discovery and for the explanation of its role in combustion.

MRS. B.
There are some circumstances which prevent the nitrogen gas, thus obtained, from being perfectly pure; but we may easily try whether the oxygen has disappeared by putting another lighted taper under it.- You see how instantaneously the flame is extinguished for want of oxygen; and were you to put an animal under the receiver, it would immediately be suffocated. But that is an experiment which, I suppose, your curiosity will not tempt you to try[1].

EMILY
It must be very cruel indeed!

The collection of oxygen and the combustion of iron[2].

EMILY
The bubbles of oxygen gas rise, I suppose, from their specific levity?

MRS. B.
Yes; for though oxygen forms rather a heavy gas, it is light compared to water. You see how it gradually replaces water from the receiver. It is now full of gas, and I may leave it inverted in water on this shelf, where I can keep the gas as long as I choose, for future experiments. This apparatus (which is indispensable for all experiments in which gasses are concerned) is called a water-bath.

---

[1] I am most grateful to Julia Saunders for telling me about the poem *The Mouse's Petition*, by Anna Barbauld (1743-1825), a friend of Joseph Priestley's. The poem was found attached to a cage in which a mouse was awaiting one of Priestley's experiments on atmospheric gases. A plea for liberation, it was hailed by reviewers as an antivivisectionist statement. The author robustly replied that what was intended as the petition of mercy against justice had been misconstrued as the plea of humanity against cruelty "which could never be apprehended from the Gentleman to whom it was addressed". She added that "the poor animal would have suffered more as a victim of domestic economy than of philosophical curiosity". . See *The Poems of Anna Letitia Barbauld*, ed. W. McCarthy and E Kraft, Athens, Ga. and London (1994) and J. Saunders, Rev. English Studies, New Series, Vol.53, No.212 (2002).

[2] M1, 136-140.

CAROLINE

It is a very clever contrivance, indeed; it is equally simple and useful. How convenient the shelf is for the receiver to rest upon under water, and the holes in it for the gas to pass into the receiver! I long to make some experiments with this apparatus.

MRS. B.

I shall try your skill in that way, when you have a little more experience. I am now going to show you an experiment, which proves, in a very striking manner, how essential oxygen is to combustion. You will see that iron itself will burn in this gas, in a most rapid and brilliant manner.

EMILY

Really! I did not know that it was possible to burn iron.

MRS. B.

Iron is eminently combustible in pure oxygen gas, and what will surprise you still more, it can be set on fire without any great rise in temperature. You see this spiral wire –I will fasten it at one end to this cork, which is made to fit an opening at the top of the glass receiver (PLATE V. Fig. 10.)[1] –

EMILY

I see the opening in the receiver; but it is carefully closed by a ground glass stopper.

MRS. B.

That is in order to prevent the gas from escaping; but I shall take out the stopper, and put in the cork, to which the wire hangs. – Now I mean to burn this wire in the oxygen gas, but I must fix a small piece of lighted tinder to the extremity of it, in order to give the first impulse to combustion; for however powerful oxygen is in promoting combustion, you must recollect that it cannot take place without a certain elevation of temperature. I shall now introduce the wire into the receiver, by quickly changing the stoppers.

CAROLINE

Is there no danger of the gas escaping while you change the stoppers?

---

[1] See p. 32.

MRS. B.

Oxygen gas is a little heavier than atmospherical air, therefore it will not mix with it very rapidly; and if I do not leave the opening uncovered, we shall not lose any –

CAROLINE

Oh, what a brilliant and beautiful flame!

EMILY

It is as white, and dazzling as the sun! – Now a piece of melted wire drops to the bottom: I fear it is extinguished; but no, it burns again as bright as ever.

MRS. B.

It will burn until the wire is entirely consumed, provided that the oxygen is not first expended; for you know that it can burn only while there is oxygen to combine with it.

CAROLINE

I never saw a more beautiful light. My eyes can hardly bear it! How astonishing to think that all this caloric was contained in the small quantity of gas that was enclosed in the receiver; and that, without producing any sensible heat!

MRS. B.

The caloric of oxygen gas could not produce any sensible heat before the combustion took place, because it was not in a free state. You can tell me, I hope, to what modification of heat this caloric is referred?

CAROLINE

Since it is *combined* with the basis of the gas, it must be *chemical* heat.

Exhaled air[1]

MRS. B

I shall show you another method of decomposing the atmosphere, which is very simple. In breathing, we retain a portion of the oxygen, and expire the nitrogen gas; so that if we breathe in a closed vessel, for a certain length of time, we shall fill it with nitrogen gas. Which of you will make the experiment?

CAROLINE
I should be very glad to try it.

MRS. B.
Very well; breathe several times through this glass tube into the receiver with which it is connected, until you feel that your breath is exhausted –

CAROLINE
I am quite out of breath already!

MRS. B.
Now let us try the gas with a lighted taper –

EMILY
It is very pure nitrogen gas, for the taper is immediately extinguished.

MRS. B.
That is not a proof of its being pure, but only of the absence of oxygen, as it is that principle alone that can produce combustion, every other gas being absolutely incapable of it.

The insignificance of nitrogen[2]

MRS. B.
I shall say nothing more of oxygen and nitrogen at present, as we shall continually have occasion to refer to the in our future conversations. They are both very abundant in nature; Nitrogen is the most plentiful in the

---

[1] M1, 145-146.
[2] M1, 148.

atmosphere, and exists also in all animal substances; oxygen forms a constituent part, both of the animal and vegetable kingdoms, from which it may be obtained by a number of chemical means. But it is now time to conclude our lesson. I am afraid you have learnt more to-day than you will be able to remember.

CAROLINE
I assure you that I have been too much interested in it, ever to forget it; as for nitrogen there seems to be but little to remember about it; it makes a very insignificant figure in comparison to oxygen, although it composes a much larger portion of the atmosphere.

MRS. B.
It will not appear so insignificant to you when you are better acquainted with it; for though it seems to perform but a passive part in the atmosphere, and has no very striking properties, when considered in its separate state, you will see by and by what a very important agent it becomes, when combined with other bodies. But no more of this at present; we must reserve it for its proper place.

CONVERSATION VI

ON HYDROGEN

Hydrogen gas as generator of water[1]

CAROLINE
The next simple body we come to is hydrogen. Pray what kind of substance is that; is it also invisible?

MRS. B.
Yes; we cannot obtain hydrogen in its pure concrete state. We are acquainted with it only in its gaseous form, as we are with oxygen and nitrogen.

CAROLINE
But in its gaseous state it cannot be called a simple substance, since it is combined with caloric.

MRS. B.
True, my dear; but as we do not know in nature of any substance which is not more or less combined with caloric, we are apt to say (rather incorrectly indeed) that a substance is in its pure state when combined with caloric only.

Hydrogen is derived from two Greek words, the meaning of which is *to generate water.*

EMILY
And how does hydrogen generate water?

---

[1] M1, 149-153.

MRS. B.
Water is composed of ... oxygen, chemically combined with.... hydrogen gas, or (as it was formerly called) inflammable air.

CAROLINE
Really! Is it possible that water should be nothing more than a combination of two gasses, and that hydrogen should be the generator of water, and at the same time inflammable air? It must be the most extraordinary gas, that will produce both fire and water!

MRS. B.
Hydrogen, I assure you, though a constituent part of water, is one of the most combustible substances in nature.

EMILY
But I thought you said that combustion could take place in no gas but oxygen?

MRS. B.
Do you recollect what the process of combustion consists in?

EMILY
In the combination of a body with oxygen, with disengagement of light and heat.

MRS. B.
Therefore, when I say that hydrogen is combustible, I mean that it has an affinity for oxygen; but, like all other combustible substances, it cannot burn unless supplied with oxygen, and heated ……….. At a certain temperature, oxygen will abandon its caloric, to combine with hydrogen; if, therefore, we raise it to that temperature, the oxygen will combine with the hydrogen, and set its own caloric at liberty; and it is thus that the combination of hydrogen gas produces water.

CAROLINE
You love to deal in paradoxes to-day, Mrs. B.– fire then produces water!

MRS. B.

The combustion of hydrogen gas certainly does; but you do not seem to have remembered the theory of combustion so well as you thought you would.

Water as an oxyd[1]

CAROLINE

Hydrogen, I see, is like nitrogen, a poor dependant friend of oxygen, which is continually forsaken for greater favourites.

MRS. B.

The connection, or friendship as you choose to call it, is much more intimate between oxygen and hydrogen, in the state of water, than between oxygen and nitrogen, in the atmosphere: for, in the first case, there is a chemical union and condensation of the two substances; in the latter, they are simply mixed together in their gaseous state……..

EMILY

Water, then, is an oxyd, though the atmospheric air is not?

MRS. B.

It is not commonly called an oxyd, though it properly belongs to that class of bodies.

---

[1] M1, 155-156.

The decomposition of water[1]

CAROLINE
I should like extremely to see water decomposed.

MRS. B.
I can easily gratify your curiosity by a much more easy process than the oxidation of charcoal or metals: the decomposition of water by these latter means, takes up a great deal of time, and is attended by much trouble; for it is necessary that the charcoal or metal should be made red hot in a furnace, that the water should pass over them in the form of vapour, that the gas be collected over the water-bath &c. In short, it is a very complicated affair. But the same effect may be produced with the greatest facility by adding some sulphuric acid (a substance with the nature of which you are not yet acquainted), to the water which the metal is to decompose………..

CAROLINE
But what metal is it that you employ for this purpose?

MRS. B.
It is iron: and it is used in the state of filings, as these present a greater surface to the acid than a solid piece of metal……… The bubbles which are now rising are hydrogen gas –

CAROLINE
How disagreeably it smells!

MRS. B.
It is indeed unpleasant, but not unwholesome[2]. We shall not, however, suffer any more to escape, as it will be wanted for experiments.

---

[1] M1, 156-158.
[2] Hydrogen itself is odourless. The smell of rotten eggs which accompanies this experiment is due to dihydrogen sulphide, formed by the action of acid on sulphide impurities in the iron. Certainly unpleasant, it is now known also to be extremely toxic.

The combustion of hydrogen gas[1]

MRS. B.

If we now set the hydrogen gas, which is contained in this receiver, at liberty all at once, and kindle it as soon as it comes in contact with the atmosphere, by presenting it to a candle, it will so suddenly and rapidly decompose the oxygen gas, by combining with its basis, that an explosion, or a *detonation* (as chemists commonly call it), will be produced. For this purpose, I need only take up the receiver, and quickly present its open mouth to the candle — so .....

CAROLINE

It produced only a sort of hissing noise, with a vivid flash of light. I had expected a much greater report.

MRS. B.

And so it would have been, had all the gasses been closely confined at the moment they were made to explode. If, for instance, we were to put in this bottle a mixture of hydrogen gas and atmospheric air; and if, after corking the bottle, we should kindle the mixture by a very small orifice, the sudden dilation of the gasses at the moment of their combination, the bottle must either fly to pieces, or the cork be blown out with considerable violence.

CAROLINE

But in the experiment which we have just seen, if you did not kindle the hydrogen gas, would it not equally combine with the oxygen?

MRS. B.

Certainly not; have I not just explained to you the necessity of the oxygen and hydrogen gasses being burnt together, in order to combine chemically and produce water?

CAROLINE

That is true; but I thought it was a different combination, for I see no water produced.

---

[1] M1, 160-167. **NO** attempt should be made to repeat **ANY** experiments involving the combustion of hydrogen.

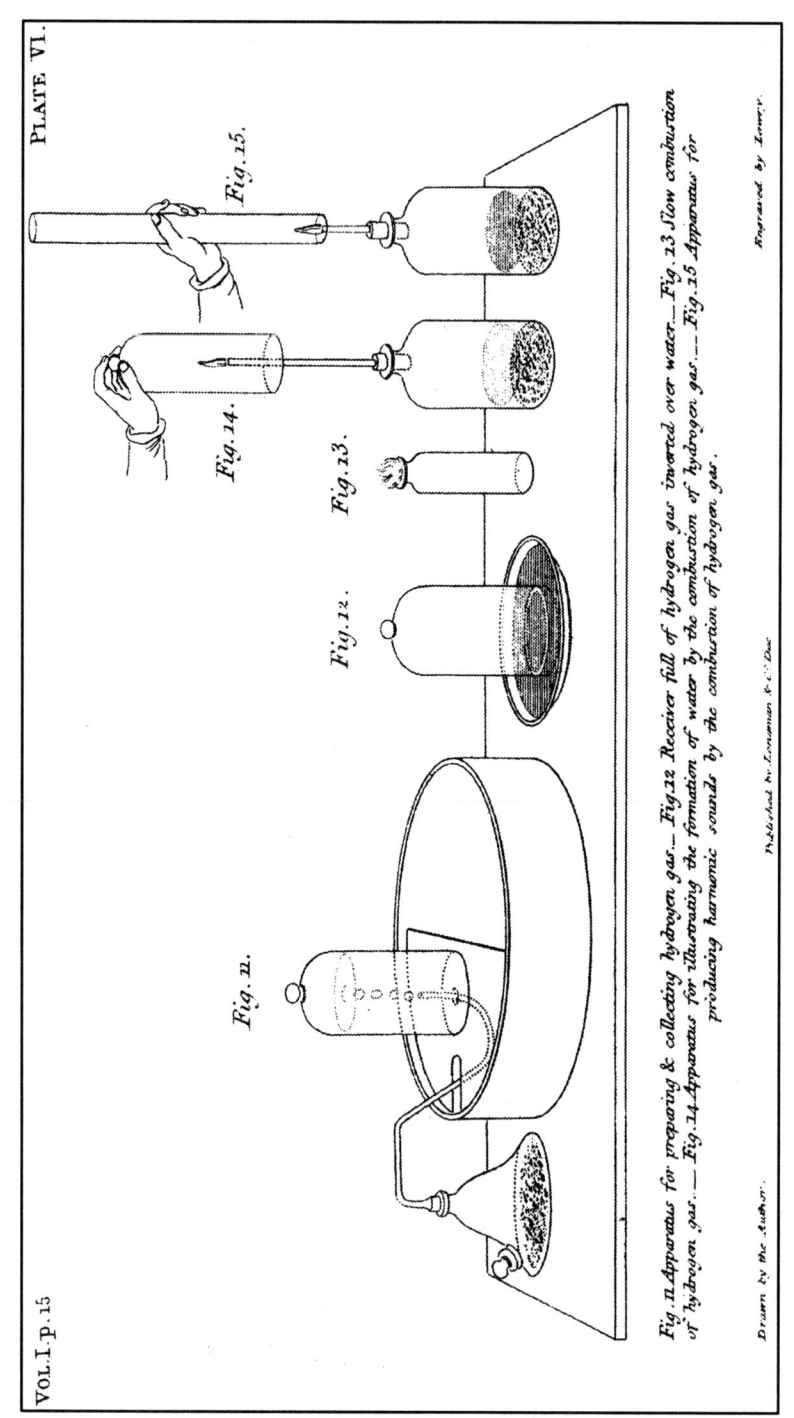

Fig. 11. Apparatus for preparing & collecting hydrogen gas.— Fig. 12. Receiver full of hydrogen gas inverted over water.— Fig. 13. Slow combustion of hydrogen gas.— Fig. 14. Apparatus for illustrating the formation of water by the combustion of hydrogen gas.— Fig. 15. Apparatus for producing harmonic sounds by the combustion of hydrogen gas.

Jane Marcet's PLATE VI

MRS. B.
The water produced by this detonation was so small in quantity, and in such a state of minute division, as to be invisible. But water certainly was produced; ………..

If, instead of bringing the hydrogen gas into sudden contact with the atmosphere (as we did just now) so as to make the whole of it explode the moment it is kindled, we allow but a very small surface of gas to burn in contact with the atmosphere, the combustion goes quietly and gradually at the point of contact, without any detonation, because the surfaces brought together are too small for the immediate union of the gasses. The experiment is a very easy one. This phial with a narrow neck (PLATE VI. Fig. 13.) is full of hydrogen gas, and is carefully corked. If I take out the cork, without moving the phial, and quickly approach the candle to the orifice, you will see how different the result will be —

EMILY
How prettily if burns, with a blue flame! The flame is gradually sinking within the phial — now it has entirely disappeared. But does not this combustion likewise produce water.

MRS. B.
Undoubtedly. In order to make the formation of water sensible to you, I shall procure a fresh supply of hydrogen gas, by putting into this bottle (PLATE VI Fig.14) iron filings, water and sulphuric acid, materials similar to those which we have just used for the same purpose. I shall then cork up the bottle, leaving only a small orifice for the cork, with a piece of glass tube fixed to it, through which the gas will issue in a continued rapid stream.

CAROLINE
I hear already the hissing of the gas through the tube, and I can feel a strong current against my hand.

MRS. B.
This current I am going to kindle with the candle — see how vividly it burns —

EMILY
It burns like a candle with a long flame. — But why does this combustion last so much longer than in the former experiment?

MRS. B.

The combustion goes on uninterruptedly as long as the new gas continues to be produced. Now if I invert this receiver over the flame, you will soon perceive its internal surface covered with a very fine dew, which is pure water —

CAROLINE

Yes, indeed; the glass is now quite dim with moisture! How glad I am that we can *see* the water produced by this combustion.

EMILY

It is exactly what I was anxious to see; for I confess that I was a little incredulous.

MRS. B.

If I had not held the glass-bell over the flame, the water would have escaped in the state of vapour, as it did in the former experiment. We have here, of course, obtained but a very small quantity of water; but the difficulty of procuring a proper apparatus, with sufficient quantities of gasses, prevents my showing it you on a larger scale.

The composition of water was discovered about the same period, both by Mr. Cavendish[1], in this country, and by the celebrated French chemist, Lavoisier. The latter invented a very perfect and ingenious apparatus to perform, with great accuracy, and upon a large scale, the formation of water by the combination of oxygen and hydrogen gasses. Two tubes, conveying due proportions, the one of oxygen, the other of hydrogen gas, are inserted at opposite sides of a large globe of glass, previously exhausted of air; the two streams of gas are kindled within the globe, by the electric spark, at the point where they come in contact; they burn together, that is to say, the hydrogen gas combines with the basis of the oxygen gas, the caloric of which is set at liberty; and a quantity of water is produced, exactly equal in weight to that of the two gasses introduced into the globe.

CAROLINE

And what was the greatest quantity of water ever formed in this apparatus?

---

[1] In 1781, Priestley and Waltire had found that a dew was formed when what we now call hydrogen was sparked with air; and in 1784, Henry Cavendish (1731-1810) showed that 80% of the air was unaffected by this reaction.

MRS. B.

Several ounces; indeed, very near a pound, if I recollect right; but the operation lasted many days.

EMILY

This experiment must have convinced all the world of the truth of the discovery. Pray, if improper proportions of the gasses were mixed and set fire to, what would be the result?

MRS. B.

Water would equally be formed, but there would be a residue of either one or other of the gasses, because, as I have already told you, hydrogen and oxygen will combine only in the proportions requisite for the formation of water.

There is another curious effect produced by the combustion of hydrogen gas, which I shall show you, though I must acquaint you first, that I cannot explain the cause of it. For this purpose, I must put some more materials into our apparatus, in order to obtain a stream of hydrogen gas, just as we have done before. The process is already going on, and the gas is rushing through the tube — I shall now kindle it with a taper —

EMILY

It burns exactly as it did before — . What is the curious effect which you were mentioning?

MRS. B.

Instead of the receiver, by means of which we have just seen the drops of water form, we shall invert over the flame this piece of tube, which is about two feet in length, and one inch in diameter (PLATE VI. Fig. 15.)[1]; but you must observe that it is open at both ends.

EMILY

What a strange noise it makes! Something like the Aeolian harp[2], but not so sweet.

CAROLINE

It is very singular, indeed; but I think rather too powerful to be pleasing. And is not this sound accounted for?[1]

---

[1] See p.44.
[2] An Aeolian harp is a sound box with tuned strings, emitting harmonics in a current of air.

MRS. B.

That the percussion of glass, by a rapid stream of gas, should produce a sound, is not extraordinary; but the sound here is so peculiar, that no other gas has a similar effect. Perhaps it is owing to the brisk vibratory motion of the glass, occasioned by the successive formation and condensation of small drops of water on the sides of the glass tube, and the air rushing in to replace the vacuum formed[2].

CAROLINE

How very much this [hydrogen] flame resembles the burning of a candle.

MRS. B.

The burning of a candle is produced by much the same means. A great deal of hydrogen is contained in candles, whether of tallow or wax. This hydrogen being converted into gas by the heat of the candle combines with the oxygen of the atmosphere, and flame and water result from this combination. So that, in fact, the flame of a candle is nothing but the combustion of hydrogen gas. An elevation of temperature, such as is produced by a lighted match or taper, is required to give the first impulse to the combustion; but afterwards it goes on of itself, because the candle finds a supply of caloric in the successive quantities of *chemical* heat which will become *sensible* by the combustion of the two gasses.

Hydrogen in soap bubbles[3]

MRS. B.

I have another experiment to show you with hydrogen gas, which I think will entertain you. Have you ever blown bubbles with soap and water?

---

[1] Now known as a *singing flame*, the sound is thought to arise from resonance between two sets of vibrations: that of the flame on its burner, and that of the outer tube. See Gaydon, A. G., and Wolfhard, H. G., *Flames*, Chapman and Hall, London, 4th edn., 1979, p.182-5.

[2] *This ingenious explanation was first suggested by Dr. Delarive.*– See *Journals of the Royal Institution, vol.i. p. 259*. This footnote is Jane Marcet's (M1, 167); Delarive was probably a medical student with Alexander Marcet at Edinburgh, where they were taught chemistry by Joseph Black.

[3] M1, 170-174.

EMILY
Yes, often. When I was a child; and I used to make them float in air by blowing them upwards.

MRS. B.
We shall fill some bubbles with hydrogen gas, instead of atmospheric air, and you will see with what ease and rapidity they will ascend, without the assistance of blowing, from the lightness of the gas — . Will you mix some soap and water whilst I fill this bladder with the gas contained in the receiver which stands on the shelf in the water-bath.

CAROLINE
What is the use of the brass stopper and turncock at the top of the receiver?

MRS. B.
It is to afford a passage to the gas when required. There is, you see, a similar stop-cock fastened to this bladder, which is made to fit that on the receiver. I screw them one on the other, and now turn the two cocks, to open a communication between the receiver and the bladder; then, by sliding the receiver off the shelf, and gently sinking it into the bath, the water rises in the receiver and forces the gas into the bladder (PLATE VII. Fig, 16.).

CAROLINE
Yes, I see the bladder swell as the water rises in the receiver.

MRS. B.
I think that we already have a sufficient quantity in the bladder for our purpose; we must be careful to stop both the cocks before we separate the bladder from the receiver, lest the gas should escape.— Now I must fix a pipe to the stopper of the bladder and, by dipping its mouth into the soap and water, take up a few drops — the I again turn the cock and squeeze the bladder in order to force the gas into the soap and water at the mouth of the pipe (PLATE VII. Fig. 17).

EMILY
There is a bubble — but it bursts before it leaves the mouth of the pipe.

MRS. B.
We must have patience and try again; it is not so easy to blow bubbles by means of a bladder, as simply with the breath.

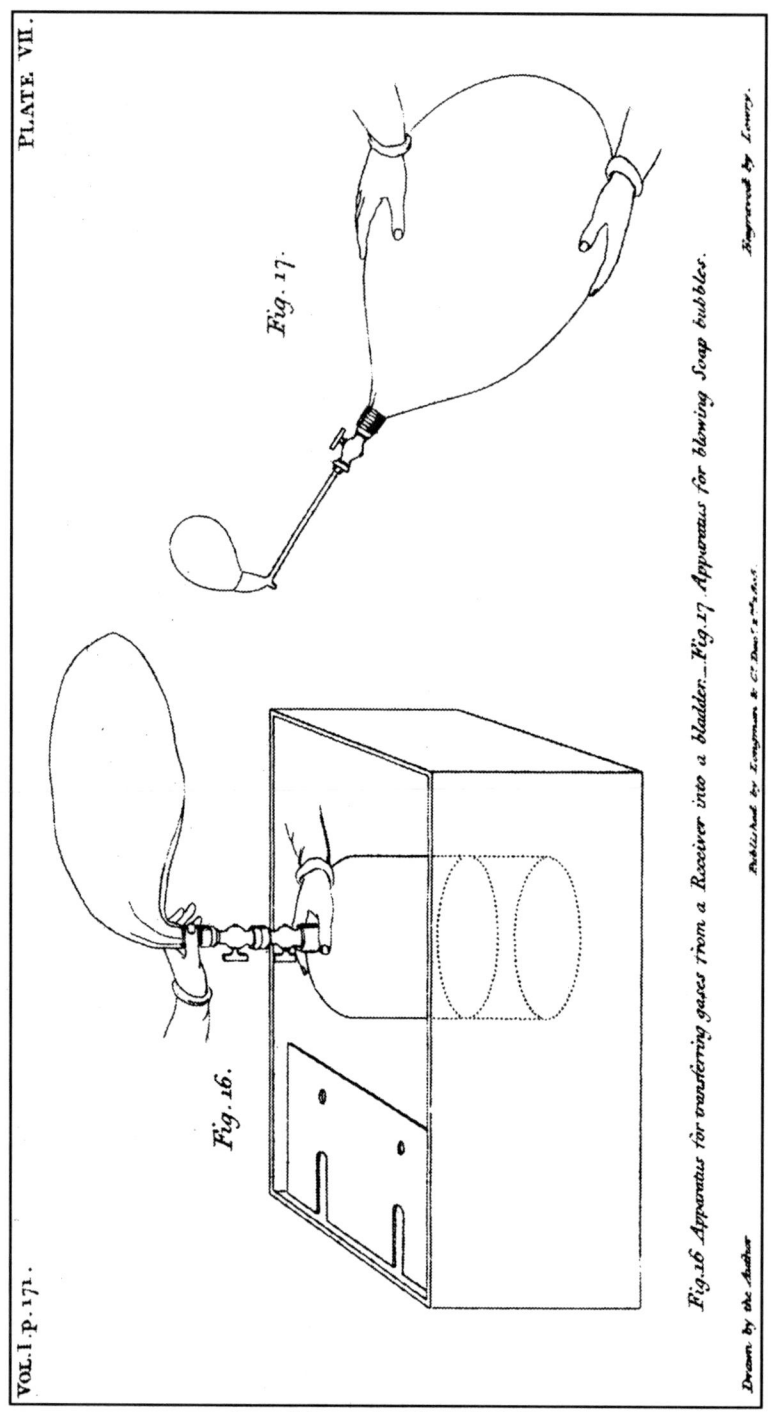

Jane Marcet's PLATE VII

CAROLINE
Perhaps there was not enough soap in the water; we should have had warm water, it would have dissolved the soap better.

EMILY
Does not some of the gas escape between the bladder and the pipe?

MRS. B.
No, they are perfectly air-tight; we shall succeed presently, I dare say.

CAROLINE
Now a bubble ascends; it moves with the rapidity of a balloon. How beautifully it refracts[1] the light!

EMILY
It has burst against the ceiling — you succeed now wonderfully; but why do they all ascend and burst against the ceiling?

MRS. B.
Hydrogen gas is so much lighter than atmospherical air, that it ascends rapidly with its very light envelop, which is burst by the force with which it strikes the ceiling.

If you are not yet tired of experiments, I have another to show you. It consists in filling soap bubbles with a mixture of hydrogen and oxygen gasses, in the proportions that form water; and afterwards setting fire to them.

EMILY
They will detonate, I suppose?

MRS. B.
Yes, they will. As you have seen the method of transferring the gas from the receiver into the bladder, it is not necessary to repeat it. I have therefore provided a bladder which contains a due proportion of oxygen and hydrogen gasses, and we have only to blow bubbles with it.

CAROLINE
Here is a fine large bubble rising — shall I set fire to it with the candle?

---

[1] The colours of a soap bubble are not, like those of a dew drop, produced by refraction, but by destructive interference of light reflected from the inner and outer surfaces of the film of liquid.

MRS. B.
If you please…

CAROLINE
Heavens[1], what an explosion!— It was like the report of a gun: I confess it frightened me much. I should never have imagined it could be so loud.

EMILY
And the flash was as vivid as lightning.

MRS. B.
The combination of the two gasses takes place during that instant of time that you see the flash and hear the detonation.

---

[1] The exclamation *Heavens* was deleted in some later editions, in response to the comments of Joseph Gurley, a schoolmaster who wrote to Jane Marcet complaining that "expressions of an irreverent character" were bad for "the youthful reader, or indeed any reader of whatever age or class", were unsuitable for young ladies, or indeed for any youthful reader. Towards the end of Jane Marcet's life, however, such expressions could again be found in her books (see also p. 129).

CONVERSATION VII

SULPHUR AND PHOSPHORUS

The sublimation of sulphur[1]

MRS. B.
Sulphur is the next simple substance that comes under our consideration. It differs in one essential point from the preceding, as it exists in a solid form at the temperature of the atmosphere.

CAROLINE
I am glad that we have at last a solid body to examine; one that we can see and touch. Pray, is it not with sulphur that the points of matches are covered to make them easily kindle?

MRS. B.
Yes, it is, and you therefore already know that sulphur is a very combustible substance. It is seldom discovered in nature in a pure unmixed state; so great is its affinity for other substances, that it is almost constantly found combined with some of them. It is most commonly united with metals, under various forms, and is separated from them by a very simple process. It exists likewise in many mineral waters, and some vegetables yield it in various proportions, especially those of the cruciform tribe. It is also found in animal matter; in short, it exists in greater or less quantity, in the mineral, vegetable and animal kingdoms.

EMILY
I have heard of *flowers of sulphur*, are they the produce of any plant?

MRS. B.
By no means: they consist of nothing more than common sulphur reduced to a very fine powder by a process known as *sublimation.*— You see some of it in this phial; it is exactly the same substance, as this lump of

---

[1] M1, 178-182.

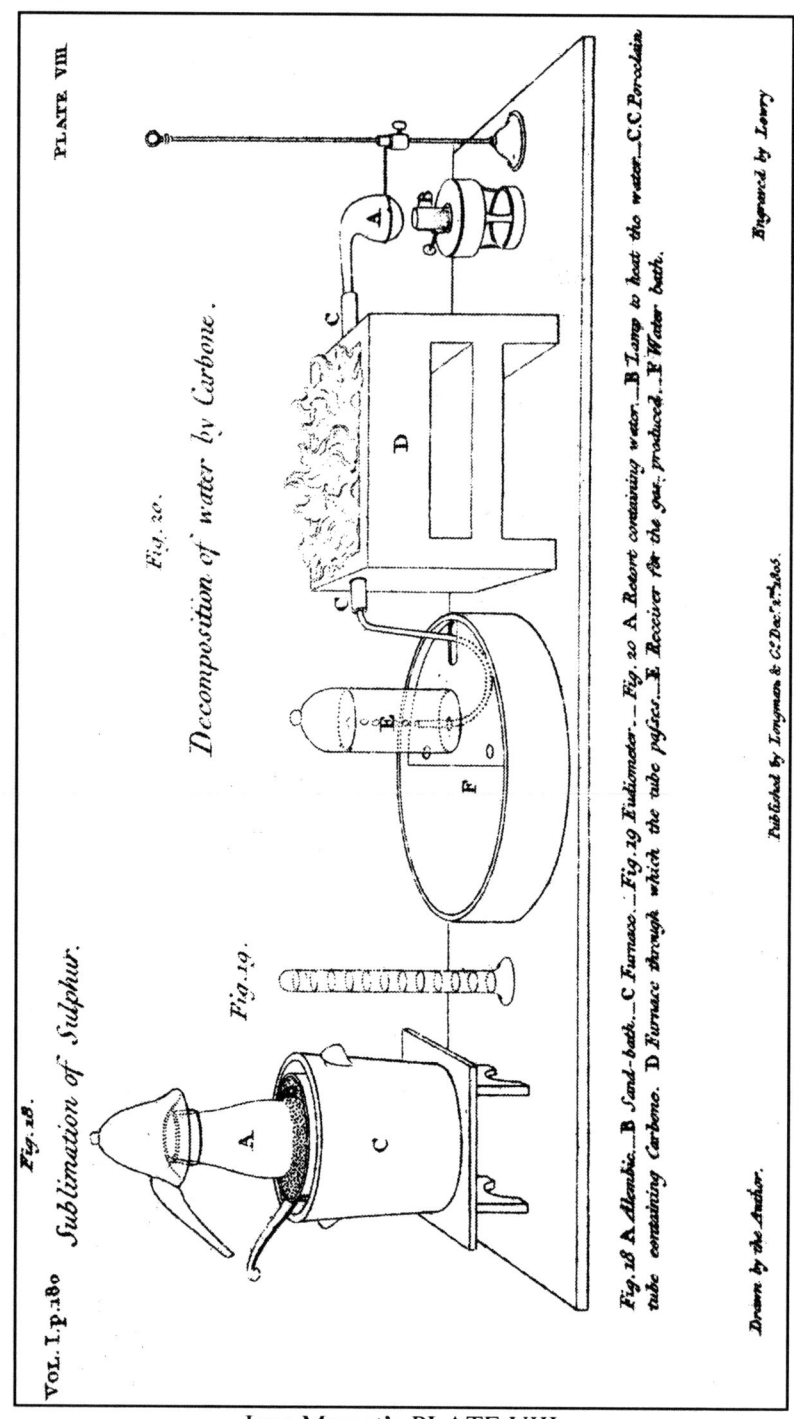

Jane Marcet's PLATE VIII

sulphur, only its colour is paler yellow, owing to it state of very minute division.

EMILY
Pray what is sublimation?

MRS. B.
It is the evaporation, or, more properly speaking, the volatilization of solid substances, which, in cooling, condense again in a concrete form. The process, in this instance, must be performed in a closed vessel, both to prevent combustion, which would take place if the access of air were not carefully precluded, and likewise to collect the substance after the operation. As it is rather a slow process, we shall not try the experiment now; but you will understand perfectly if I show you the apparatus used for the purpose.– (PLATE VIII, Fig. 18). Some lumps of sulphur are put into a receiver of this kind, called a *cucurbit*. Its shape, you see somewhat resembles that of a pear, and it is open at the top so as to adapt itself exactly to a kind of conical receiver of this sort called the head. The cucurbit, thus covered with its head is placed over a sand-bath; this is nothing more than a vessel full of sand, which is kept heated by a furnace, such as you see here, so as to preserve the apparatus in a moderate and uniform temperature. The sulphur then soon begins to melt, and immediately after this, a thick white smoke rises, which is gradually deposited within the head, or upper part of the apparatus, where it condenses against the sides, somewhat in the form of a vegetation, whence it has obtained the name flowers of sulphur.....

... You now perfectly understand, I suppose, what is meant by sublimation?

EMILY
I believe I do. Sublimation appears to consist in destroying, by means of heat, the attraction of aggregation of the particles of a solid body, which are thus volatilized; and as soon as they lose the caloric which produced that effect, they are deposited in the form of a fine powder.

CAROLINE
It seems to me to be somewhat similar to the transformation of water into vapour, which returns to its liquid state when deprived of caloric.

EMILY
There is this difference. However, that the sulphur does not return to its former state, since, instead of lumps, it changes to a fine powder.

MRS. B.

Chemically speaking, it is exactly the same substance, whether in the form of lump or powder. For if this powder be melted again by heat, and then suffered to cool, it will be restored to the same solid state as it was before sublimation.

CAROLINE

But if there be no real change, produced by the sublimation of the sulphur, what is the use of the operation?

MRS. B.

It divides the sulphur into very minute particles, and thus disposes it to enter more readily into combination with other bodies. It is used also as a means of purification.

The combustion of sulphur[1]

CAROLINE

Sublimation appears to me like the beginning of a combustion, for the completion of which one circumstance only is wanting, the absorption of oxygen.

MRS. B.

But that circumstance is every thing. No essential alteration is produced in sulphur by sublimation; whilst in combustion it combines with the oxygen and forms a new compound totally different in every respect from sulphur in its pure state.— We shall now *burn* some sulphur, and you will see how very different the result will be. For this purpose I put a small quantity of flowers of sulphur into this cup, and place it in a dish, into which I have poured a little water; I now set fire to the sulphur with the point of this hot wire; for its combustion will not begin until its temperature be considerably raised.— You see it begins to burn with a faint blueish flame; and as I invert over it this receiver, white fumes arise from the sulphur and fill the receiver.— You will soon perceive that the water is rising within the receiver, a little above its level in the plate.— Well, Emily, can you account for this?

---

[1] M1, 182-184.

EMILY

I suppose that the sulphur has absorbed the oxygen from the atmospherical air within the receiver; and that we shall find some oxygenated sulphur in the cup. As for the white smoke, I am quite at a loss to guess what it might be.

MRS. B.

Your first conjecture is very right; but you are quite mistaken in the last; for nothing will be left in the cup. The white vapour is oxygenated sulphur, which assumes the form of an elastic fluid of a pungent and offensive smell, and is a powerful acid. Here you see a chemical combination of oxygen and sulphur, producing a true gas, which would continue such under the pressure and at the temperature of the atmosphere, if it did not unite with the water in the plate, to which it imparts its acid taste and all its acid properties.— You see, now, with what curious effects the combustion of sulphur is attended.

CAROLINE

This is something quite new; and I confess that I do not perfectly understand why the sulphur turns acid.

MRS. B.

It is because it unites with oxygen, which is the general acidifying principle. And, indeed, the word *oxygen*, is derived from two Greek words signifying *to generate an acid*[1].

Acids[2]

MRS. B.

I believe it will be necessary, before we proceed further, to say a few words of the general nature of acids, though it is rather a deviation from our plan of examining the simple bodies separately, before we consider them in a state of combination.

Acids may be considered as a particular class of *burnt* bodies, which during their combustion, or combination with oxygen, have acquired very

---

[1] It was unwise of Lavoisier to assume that all acidic substances, whether known or yet to be discovered, would be shown to contain "oxygen". (See pp. 107, 148).
[2] M1, 188-191.

characteristic properties. They are chiefly discernable by their sour taste, and by turning red most of the blue vegetable colours. These two properties are common to the whole class of acids; but each of them is distinguished by other peculiar qualities. Every acid consists of some particular substance (which constitutes its basis, and is different in each), and of oxygen, which is common to them all.

EMILY
But I do not clearly see the difference between oxyds and acids?

MRS. B.
Acids were, in fact, oxyds, which, by the addition of a sufficient quantity of oxygen, have been converted into acids. For acidification, you must observe, always implies previous oxydation, as the body must have combined with the quantity of oxygen requisite to constitute an oxyd, before it can combine with the greater quantity that is necessary to render it an acid.

CAROLINE
Are all oxyds capable of being converted into acids?

MRS. B.
Very far from it; it is only certain substances which will enter into that peculiar kind of union with oxygen which produces acids, and the number of these is proportionally very small; but all burnt bodies may be considered as belonging either to the class of oxyds, or to that of acids. At a future period, we shall enter more at large into this subject. At present, I have but one circumstance further to point out to your observation respecting acids; it is, that most of them are susceptible of two degrees of acidification, according to the different quantities of oxygen with which their basis combines.

EMILY
And how are these two degrees of acidification distinguished?

MRS. B.
By the peculiar properties that result from them. The acid we have just made is the first or weakest degree of acidification, and is called *sulphureous acid*; if it were fully saturated with oxygen, it would be called *sulphuric acid*. You must therefore remember, that in this, as in all acids, the first degree of acidification is expressed by the termination in *ous*; the stronger, by the termination in *ic*.

Sulphurated hydrogen gas[1]

MRS. B.
Before we quit the subject the subject of sulphur, I must tell you that it is susceptible of combining with a great variety of substances, and especially with hydrogen, with which you are already acquainted. Hydrogen gas can dissolve a portion of it.

EMILY
What; can a gas dissolve a solid substance?

MRS. B.
Yes; a solid substance may be so minutely divided by heat, as to become soluble in a gas; and there are several instances of it. But you must observe that, in this case, a chemical solution, that is to say, a combination of the sulphur with the hydrogen gas, is produced. In order to effect this, the sulphur must be strongly heated in contact with the gas; the heat reduces the sulphur to such a state of extreme division, and diffuses so strongly through the gas, that they combine and incorporate together. And as a proof that there must be a chemical union between the sulphur and the gas, it is sufficient to remark that they are not separated when the sulphur loses the caloric be which it was volatilized. Besides, it is evident, from the peculiar fetid smell of the gas, that it is a new compound totally different from either of its constituents; it is called *sulphurated hydrogen gas*, and is contained in great abundance in sulphureous mineral waters.

CAROLINE
Are not the Harrogate waters of this nature?

MRS. B.
Yes; they are naturally impregnated with sulphurated hydrogen gas, and there are many other springs of the same kind; which shows that this gas must often be formed in the bowels of the earth by spontaneous processes of nature.

---

[1] M1, 188-91.

CAROLINE
And could not such waters be made artificially by impregnating common water with this gas?

MRS. B.
Yes; they can be so well imitated as perfectly to resemble the Harrogate waters.

Sulphur combines likewise with phosphorus, and with the alkalies, and alkaline earths, substances with which you are yet unacquainted. We cannot, therefore, enter into these combinations at present. In our next lesson we shall treat phosphorus.

EMILY
May we not begin that subject to-day; this lesson has been so short?

MRS. B.
I have no objection, if you are not tired. What do you say, Caroline?

CAROLINE
I am as desirous as Emily of prolonging the lesson to-day, especially as we are to enter on a new subject; for I must confess that sulphur has not appeared so interesting to me as the other simple bodies.

MRS. B.
Perhaps you may find phosphorus more entertaining. You must not, however, be discouraged when you meet some parts of a study less amusing than others; it would answer to no good purpose to select the most pleasing parts, since, if we did not proceed with some method, in order to acquire a general idea of the whole, we could scarcely expect to take any interest in particular subjects.

The discovery of phosphorus[1]

MRS. B.
Phosphorus is a simple substance that was formerly unknown. It was first discovered by Brandt, a chemist of Hamburgh, whilst employed in researches after the philosopher's stone; but the method of obtaining it

---

[1] M1, 191-193. A lively account of these events is given by John Emsley in *The Shocking History of Phosphorus*, Macmillan, London (2000).

remained a secret till it was a second time discovered, both by Kunckel and Boyle, in the year 1680............

CAROLINE

I do not understand in what the discovery consisted; there may be a secret method of making a composition, but a simple body cannot be *made*, it can only be *found*.

MRS. B.

But a body may exist in nature so closely combined with other substances, as to elude the observations of chemists, or render it extremely difficult to obtain it in its simple state. This is the case with phosphorus, which is always so intimately combined with other substances, that its existence remained unnoticed until Brandt discovered the means of obtaining it free from all combinations. It is found in all animal substances, and is now chiefly extracted from bones, by a chemical process. It exists also in some plants, that bear a strong analogy to animal matter in their chemical composition.

EMILY

But is it never found in its simple state?

MRS. B.

Never, and this is the reason of its having remained so long undiscovered.

EMILY

Is it possible, then, that in the course of time we may discover other new simple bodies?

MRS. B.

Undoubtedly; and we may also learn that some of those, which we now class among the simple bodies, may, in fact, be compound; indeed, you will soon find that discoveries of this kind are by no means unfrequent.

The combustion of phosphorus[1]

MRS. B.
Phosphorus is eminently combustible; it melts and takes fire at the temperature[2] of 100°, and absorbs in its combustion nearly once and a half its own weight of oxygen.

CAROLINE
What! Will a pound of phosphorus consume a pound and a half of oxygen?

MRS. B.
So it appears from accurate experiments. I can show you with what violence it combines with oxygen, by burning some of it in that gas. We must manage the experiment in the same manner as we did the combustion of sulphur.— You see I am now obliged to cut this little bit of phosphorus under water, otherwise there would be danger of its taking fire by the heat of my fingers.— I now put it into the receiver, and kindle it by means of a hot wire.

EMILY
What a blaze! I can hardly bear to look at it. I never saw anything so brilliant. Does it not hurt your eyes, Caroline?

CAROLINE
Yes; but still I cannot help looking at it. A prodigious quantity of oxygen must indeed be absorbed, when so much light and caloric are disengaged!

MRS. B.
In the combustion of a pound of phosphorus, a sufficient quantity of caloric is set free to melt upwards of a hundred pounds of ice; this has been computed with direct experiments with the calorimeter[3].

EMILY
And is the result of this combustion, like that of sulphur, an acid?

---

[1] M1, 193-195, 197-198.
[2] Jane Marcet is using the Fahrenheit scale; 100°F is about 38°C, a little above human body temperature
[3] These quantities are in rough agreement with those obtained from modern data.

MRS. B.

Yes; phosphoric acid. And had we duly proportioned the phosphorus and the oxygen, they would have been completely converted into phosphoric acid, weighing together, in this new state, exactly the sum of their weights separately. The water would have ascended into the receiver, on account of the vacuum formed, and would have filled it entirely. In this case, as in the combustion of sulphur, the acid vapour formed is absorbed and condensed in the water of the receiver. But when this combustion is formed without any water or moisture being present, the acid then appears in the form of concrete whitish flakes, which are, however, extremely ready to melt upon the least admission of water.

EMILY

Does phosphorus, in burning in atmospherical air, produce, like sulphur, a weaker sort of the same acid?

MRS. B.

No: for it burns in atmospherical air nearly at the same temperature as in oxygen gas; and it is, in both cases, so strongly disposed to combine with the oxygen, that the combustion is perfect, and the product similar; only in atmospherical air, being less rapidly supplied with oxygen, the process is performed in a slower manner............

.........Phosphorus is sometimes used as a test to estimate the purity of atmospherical air. For this purpose, it is burnt in a graduated tube called an eudiometer (PLATE VIII. Fig. 19)[1], and from the quantity of air which the phosphorus absorbs, the proportion of oxygen in the air examined, is deduced; for the phosphorus will absorb all the oxygen, and the nitrogen alone will remain.

EMILY

And the more oxygen is contained in the atmosphere, the purer, I suppose, it is esteemed?

MRS. B.

Certainly.

---

[1] See p. 54.

Phosphorus and sulphur[1]

MRS. B.
Phosphorus, when melted, combines with a great variety of substances. With sulphur it forms a compound so extremely combustible, that it immediately takes fire on coming in contact with the air. It is with this composition that the phosphoric matches are prepared, which kindle as soon as they are taken out of their cases and are exposed to the air.

EMILY
I have a box of these curious matches; but I have observed, that in very cold weather, they will not take fire without being previously rubbed.

MRS. B.
By rubbing them you raise their temperature; for, you know, friction is one of the means of extricating heat.

Phosphorated hydrogen gas[2]

EMILY
Will phosphorus combine with hydrogen gas, as sulphur does?

MRS. B.
Yes; and the compound gas which results from this combination has a smell still more fetid than that of sulphurated hydrogen; it resembles that of garlic.

The *phosphorated hydrogen gas*[3] has this remarkable peculiarity, that it takes fire spontaneously in the atmosphere, at any temperature. It is thus that are produced those transient flames, or flashes of light, called by the vulgar *Will-of-the-Wisp*, or more properly *Ignes-fatui*, which are often seen

---

[1] M1, 198.
[2] M1, 198-202.
[3] The simplest compound of phosphorus and hydrogen is phosphine, PH_3, which does not catch fire spontaneously in air unless it is contaminated with its less stable analogue, biphosphine, P_2H_4. It is thought that traces of biphosphine, produced by bacterial decay of animal and vegetable matter give rise to the phenomenon of Will-o'-the-Wisp over marshy ground and also to reports of luminous "ghosts" in churchyards. Unfortunately, such sightings are nowadays rare, owing to drainage of marshes and light-pollution near towns.

in church-yards, and places where the putrefaction of animal matter exhales phosphorus and hydrogen gas.

CAROLINE
Country people, who are so much frightened by those appearances, would soon be reconciled to them, if they knew from what a simple cause they proceed.

MRS. B.
There are other combinations of phosphorus that also have very singular properties, particularly that which results from its union with lime............
...This latter compound....... has the singular property of decomposing water, merely by being thrown into it. It effects this by absorbing the oxygen of water, in consequence of which bubbles of hydrogen gas ascend, holding in solution a small quantity of phosphorus.

EMILY
These bubbles then are *phosphorated hydrogen gas?*

MRS. B.
Yes; and they produce the singular appearance of a flash of fire issuing from water, as the bubbles kindle and detonate on the surface of the water, at the instant they come in contact with the atmosphere.

CAROLINE
Is this effect nearly similar to that produced by the combination of phosphorus with sulphur, or, more properly speaking, the *phosphoret of sulphur?*

MRS. B.
Yes; but the phenomenon appears more extraordinary in this case, from the presence of water and from the gaseous form of the combustible compound. Besides, the experiment surprises by its great simplicity. You only throw a piece of phosphoret of lime into a glass of water, and bubbles of fire will immediately issue from it.

CAROLINE
Cannot we try the experiment?

MRS. B.

Very easily; but we must do it in the open air; for the smell of the phosphorated hydrogen gas is so extremely fetid, that it would be intolerable in the house. But before we leave the room, we may produce, by another process, some bubbles of phosphorated hydrogen gas, which are much less offensive.

There is in this little glass retort a solution of potash in water; I add to it a small piece of phosphorus. We must now heat the retort over a lamp, after having engaged its neck under water— you will see it begins to boil; in a few minutes bubbles will begin to appear, which take fire and detonate as they issue from the water.

CAROLINE

There is one— and another. How curious it is!— But I do not understand how this is produced?

MRS. B.

It is the consequence of a display of affinities too complicated, I fear, to be made perfectly intelligible to you at present[1].

In a few words, the reciprocal action of the potash, phosphorus, caloric and water, are such that some of the water is decomposed, and the hydrogen gas thereby formed carries off some minute particles of phosphorus, with which it forms phosphorated hydrogen gas, a compound which takes fire at almost any temperature.

---

[1] Yes, indeed. See p. 105, notes 1 and 2.

CONVERSATION VIII

ON CARBONE

The imitation of nature[1]

MRS. B.
This ring, which I wear on my finger, owes its brilliancy to a small piece of carbone.

CAROLINE
Surely, you are joking, Mrs. B.?

EMILY
I thought your ring was diamond?

MRS. B.
It is so, but diamond is nothing more than carbone in its purest and most perfect state.....

CAROLINE
But pray, Mrs. B—, since it is known of what substance diamond and cotton are composed, why should they not be manufactured, or imitated by some chemical process, which would render them much cheaper and more plentiful than the present mode of obtaining them?

MRS. B.
With our system, Caroline, you might a well propose that we should make flowers and fruit, nay, perhaps even animals, by a chemical process; for it is known of what these bodies consist, since every thing which we are acquainted with in nature, is formed from the various simple substances that we have enumerated. But, my dear, we must not suppose that a knowledge of the component parts of a body will in every case enable us to imitate it..... . The more complicated combinations of nature,

---

[1] M1, 206-209.

even in the mineral kingdom, are in general beyond our reach, and any attempt to imitate organised bodies must probably ever prove fruitless; their formation is a secret that rests in the bosom of the Creator. You see, therefore, how vain it would be to attempt the formation of cotton by chemical means. But, surely, we have no reason to regret our inability in this instance, when nature has so clearly pointed out a method of obtaining it in perfection and abundance.

CAROLINE
I did not imagine that the principle of life could be imitated by the aid of chemistry; but it did not appear to me ridiculous to suppose that chemists might attain a perfect imitation of inanimate nature…..

EMILY
But diamond, since it consists merely of one simple unorganised substance, might be, one would think, perfectly imitable by art?

MRS. B.
It is sometimes as much beyond our power to obtain a simple body in a state of perfect purity, as it is to imitate a complicated combination; for the operations by which nature decomposes bodies are frequently as inimitable as those which she uses for their combination.

Carbonic acid gas[1]

[Mrs. B. has just demonstrated the combustion of carbon in oxygen]

EMILY
The charcoal is now extinguished, although it is not nearly consumed; it has such an extraordinary avidity for oxygen, I suppose, that the receiver did not contain enough to satisfy the whole.

MRS. B.
That is certainly the case; for if the combustion was performed in the exact proportion of 28 parts[2] of carbone to 72 of oxygen, both ingredients would disappear, and 100 parts of carbonic acid would be produced.

---

[1] M1, 217-21.
[2] Jane Marcet uses *part* to mean *part by mass.*

CAROLINE
Carbonic acid must be a very strong acid, since it contains[1] so great a proportion of oxygen?

MRS. B.
That is a very natural inference; yet it is erroneous. For carbonic acid is the weakest of all acids. The strength of an acid seems to depend on the nature of its basis and its mode of combination, as well as on the proportion of the acidifying principle. The same quantity of oxygen that will convert some bodies into strong acids, will only be sufficient simply to oxydate others.

CAROLINE
Since this acid is so weak, I think chemists should have called it the *carbonous*, instead of the *carbonic* acid.

EMILY
But, I suppose, the carbonous acid is still weaker, and is formed by burning carbone in atmospherical air.

MRS. B.
No, my dear. Carbone does not seem to be susceptible of more than one degree of acidification, whether burnt in oxygen gas, or atmospherical air. There is therefore no carbonous acid.

It has indeed been lately discovered, that carbone may be converted into a gas, by uniting with a smaller proportion of oxygen; but as this gas does not possess any acidic properties, it is no more than an oxyd; and in order to distinguish it from charcoal, which contains a still smaller proportion of oxygen, it is called *gaseous oxyd of carbone*.....

CAROLINE
......Pray can this gas [*carbonic acid gas*] be condensed into a liquid?

---

[1] These figures were doubtless based on experimental values. It is no surprise that no reference is made to the "relative weights" of combining masses which formed the basis of the emergent atomic theory which was being formulated by John Dalton (1766 – 1844) in the summer of 1804. Even if Jane Marcet had known of this work, her express aim was to supplement Davy's lectures, and at first he opposed the atomic theory with public ridicule. By 1808, however, he had been converted to it, and atoms were accordingly given their due place in later editions of *Conversations on Chemistry*.

MRS. B.

No: for, as I told you before, it is a permanent elastic fluid. But water can absorb a certain quantity of this gas, and can even be impregnated with it, in a very strong degree, by the assistance of agitation and pressure, as I am going to show you. I shall decant some carbonic acid gas into this bottle, which I shall fill first with water, in order to exclude the atmospherical air; the gas is then introduced through the water, which you see it displaces, for it will not mix with any quantity unless strongly agitated, or allowed to stand over it for some time. The bottle is now about half full of carbonic acid gas, and the other half is still occupied by the water. By corking the bottle, and then violently shaking it, in this way, I can mix the gas and water together.— Now will you taste it?

EMILY

It has a distinct acid taste.

CAROLINE

Yes, it is sensibly sour, and full of little bubbles.

MRS. B.

It possesses likewise all the other properties of acids, but of course in a much less degree than the pure carbonic acid gas, as it is so much diluted with water.

This is a kind of artificial Seltzer water[1]. By analysing that which is produced by nature, it was found to contain scarcely any thing more than common water impregnated with a certain proportion of carbonic acid gas. We are, therefore, able to imitate it, by mixing those proportions of water and carbonic acid. Here, my dear, is an instance in which, by a chemical process, we can exactly copy the operations of nature; for the artificial Seltzer waters can be made in every respect similar to those of nature: in one point, indeed, the former have an advantage, as they may be prepared stronger, or weaker, as occasion requires.

CAROLINE

I thought I had tasted such water before, but what makes it so brisk and sparkling?

---

[1] Selterser-water was a sparkling natural water from the German village of Seltsers, near Frankfurt.

MRS. B.

This sparkling, or effervescence, as it is called, is always occasioned by an elastic fluid escaping from a liquid; in artificial Seltzer water it is produced by the carbonic acid, which being lighter than the water in which it was strongly condensed, flies off with great rapidity the instant the bottle is uncorked; this makes it necessary to drink it immediately. The bubbling that took place in this bottle was but trifling, as the water was but very slightly impregnated with carbonic acid. It requires a particular apparatus to prepare the gaseous artificial mineral waters.

Hot coals and water[1]

EMILY

If I recollect right, Mrs. B., you told us that carbone was capable of decomposing water; the affinity between water and carbone must therefore be greater than between oxygen and hydrogen?

MRS. B.

Yes, but this is not the case unless their temperature be raised to a certain degree. It is only when carbone is red hot, that it is capable of separating the oxygen from the hydrogen. Thus, if a small quantity of water be thrown on a red hot fire, it will increase, rather than extinguish the combustion; for the coals or wood (both of which contain a great quantity of carbone) decompose the water, and thus supply the fire both with oxygen and hydrogen gasses. If, on the contrary, a large mass of water be thrown over the fire, the diminution of heat thus produced is such that the combustible matter loses the power of decomposing the water, and the fire is extinguished.

EMILY

I have heard that fire-engines sometimes do more harm than good, and that they actually increase the fire when they cannot throw water enough to extinguish it. It must be owing, no doubt, to the decomposition of the water by the carbone during the conflagration.

---

[1] M1, 223-225.

MRS. B.

Certainly.– The apparatus which you see here (PLATE VIII. Fig. 20)[1] may be used to exemplify what we have just said. It consists of a kind of open furnace, through which a porcelain tube, containing charcoal, passes. To one end of the tube is attached a glass retort with water in it; and the other end communicates with a receiver placed on the water bath.– A lamp being applied to the retort, and the water made to boil, the vapour is gradually conveyed through the red hot charcoal, by which it is decomposed; and the hydrogen gas which results from this decomposition is collected in the receiver. But the hydrogen thus obtained is far from being pure; it retains in solution a minute portion of carbone, and contains also a quantity of carbonic acid. This renders it heavier than pure hydrogen gas, and gives it some peculiar properties: it is distinguished by the name of *carbonated hydrogen gas*.

CAROLINE

And whence does it obtain the carbonic acid that is mixed with it?

EMILY

I believe I can answer that question, Caroline.– From the union of the oxygen (proceeding from the decomposed water) with the carbone, which, you know, makes carbonic acid.

CAROLINE

True; I should have recollected that.– The product of the decomposition of water by red hot charcoal, therefore, is carbonated hydrogen gas.

MRS. B.

You are perfectly right now.

---

[1] See p. 54.

Candles and oil lamps[1]

MRS. B.
The combustion of a candle, and that of a lamp, both produce water and carbonic acid gas. Can you tell me how these are formed?

EMILY
Let me think.... Both the candle and lamp burn by means of fixed oil[2]— this is decomposed as the combustion goes on; and the constituent parts of the oil being thus separated, the carbone unites to a portion of the oxygen from the atmosphere to form carbonic acid gas, whilst the hydrogen combines with another portion of oxygen, and forms with it water.— The products, therefore of the combustion of oils, are water and carbonic acid gas.

CAROLINE
But we see neither water nor carbonic acid produced by the combustion of a candle?

MRS. B.
The carbonic acid gas, you know, is invisible, and the water being in a state of vapour, is so likewise. Emily is perfectly correct in her explanation, and I am very much pleased with it.

EMILY
I am very much delighted with all these new ideas; but, at the same time, I cannot help being apprehensive that I may forget some of them.

MRS. B.
I would advise you to take notes, or what would answer better still, to write down, after every lecture, as much of it as you can recollect. And, in order to give you a little assistance, I shall lend you the heads or index, which I occasionally consult for the sake of preserving some method and arrangement in these conversations. Unless you follow some plan, you

---

[1] M1, 231-3.
[2] from M1, 230: *MRS. B.: Fixed oils are those which will not evaporate without being decomposed; this is the case for all the common oils which contain a greater proportion of carbone than the essential oils.....(which comprehend the whole class of essences and perfumes).*

cannot expect to retain nearly all that you learn, how great soever be the impression it may make on you at first.

EMILY

I will certainly follow your advice.— Hitherto I have found that I recollected pretty well what you have taught us; but the history of carbone is a more extensive subject than any of the simple bodies we have yet examined.

CONVERSATION IX

ON METALS

Metals[1]

MRS. B.
The metals, which we are now to examine, are bodies of a very different nature from those which we have hitherto considered. They do not, like the elements of gasses, elude the immediate observation of our senses; for they are the most brilliant, the most ponderous, and the most palpable substances in nature.

CAROLINE
I doubt, however, whether the metals will appear to us so interesting, and give us so much entertainment as those mysterious elements which conceal themselves from our view. Besides, they cannot afford so much novelty; they are bodies with which we are already so well acquainted.

MRS. B.
But that acquaintance, you will soon perceive, is but very superficial; and I trust that you will find both novelty and entertainment in considering the metals in a chemical point of view. To treat this subject fully, would require a whole course of lectures; for metals form of themselves a most important branch of practical chemistry. We must, therefore, confine ourselves to a general view of them. These bodies are seldom found naturally in their metallic form; they are generally more or less oxygenated, or combined with sulphur, earths or acids, and are often blended with each other. They are found in the bowels of the earth in most parts of the globe, but chiefly in mountainous districts, where the surface of the globe has suffered from earthquakes, volcanoes and other convulsions of nature. They are there spread in strata or beds, called veins, and these veins are composed of a certain quantity of metal, combined with various earthy substances, with which they form minerals of different nature are appearance, which are called ores.......

---

[1] M1, 236-240.

......[*the*] attraction [*of metals*] for oxygen varies extremely: there are some that will combine with it only at very high temperature, or by the assistance of acids; while there are others that oxygenate of themselves very rapidly, even at the lowest temperature, as manganese, which scarcely ever exists in its metallic state, as it immediately absorbs oxygen on being exposed to the air, and crumbles to an oxyd in the course of a few hours.

EMILY

Is it not from that oxyd that you extracted oxygen gas?

MRS. B.

It is; so that, you see, this metal attracts oxygen at a low temperature, and parts with it when strongly heated.

EMILY

Is there any other metal that oxydates at the temperature of the atmosphere?

MRS. B.

They all do, more or less, excepting gold, silver, and platina.

Copper, lead, and iron oxydate slowly in air, and cover themselves with a sort of rust, which is nothing but the general conversion of the surface into an oxyd. This rusty surface preserves the interior of the metal from oxydation, as it prevents the air from coming in contact with it.

CAROLINE

So, the, what we commonly call rust, is only an oxyd of the metal?

MRS. B.

Exactly So.

The combustion of iron filings[1]

MRS. B.

I must show you some instances of the combustion of metals; it would require the heat of a furnace to make them burn in common air, but if we supply them with a stream of oxygen gas, we may easily accomplish it.

CAROLINE

But it will still, I suppose, be necessary in some degree to raise the temperature; for the oxygen will not be able to penetrate such dense substances, unless the caloric forces a passage for it.

MRS. B.

This, as you shall see, is very easily done, particularly if the experiment be tried upon a small scale.— I begin by lighting this piece of charcoal with the candle, and then increase the rapidity of its combustion by blowing upon it with a blow-pipe. (PLATE IX. Fig. 21.)[2] —

EMILY

That I do not understand; for it is not every kind of air, but merely oxygen gas, that produces combustion. Now you said that in breathing we inspired, but did not expire, oxygen gas. Why, therefore, should the air which you breathe through the blow-pipe, promote the combustion of the charcoal?

MRS. B.

Because the air, which has but once passed through the lungs, is yet but little altered, a small portion only of its oxygen being destroyed; so that a great deal more is gained by increasing the rapidity of the current, by means of the blow-pipe, than is lost in consequence of the air passing once through the lungs, as you shall see—

EMILY

Yes, indeed, it makes the charcoal burn much brighter.

MRS. B.

Whilst it is red hot, I shall drop some iron filings on it, and supply them with a current of oxygen gas, by means of this apparatus, (PLATE IX.

---

[1] M1, 243-245.
[2] See p.79.

Fig. 22) which consists simply of a closed tin cylindrical vessel, full of oxygen gas, with two apertures and stop-cocks, by one of which a stream of water is thrown into the vessel through a long funnel, whilst by the other the gas is forced out through a blow-pipe adapted to it, as the water gains admittance.— Now that I pour the water into the funnel, you may hear the gas issuing from the blow-pipe— I bring the charcoal close to the current, and pour the filings upon it—

CAROLINE
They emit much the same vivid light, as the combustion of iron wire in oxygen gas.

MRS. B.
The process is, in fact, the same; there is only some difference in the mode of conducting it. Let us burn some tin in the same manner— you see it is equally combustible— Let us now try some copper—

CAROLINE
This burns with a greenish flame; it is, I suppose, owing to the colour of the oxyd?

The decomposition of water by metals[1]

MRS. B.
Iron, zinc, tin and antimony have a stronger affinity for oxygen than hydrogen has, therefore these four metals are capable of decomposing water. But hydrogen having an advantage over all the other metals with respect to its affinity for oxygen, it not only withholds its oxygen from them, but is even capable, in certain circumstances, of taking the oxygen from the oxyd of these metals.

EMILY
I confess I do not quite understand why hydrogen can take oxygen from those metals that do not decompose water.

---

[1] M1, 249-250.

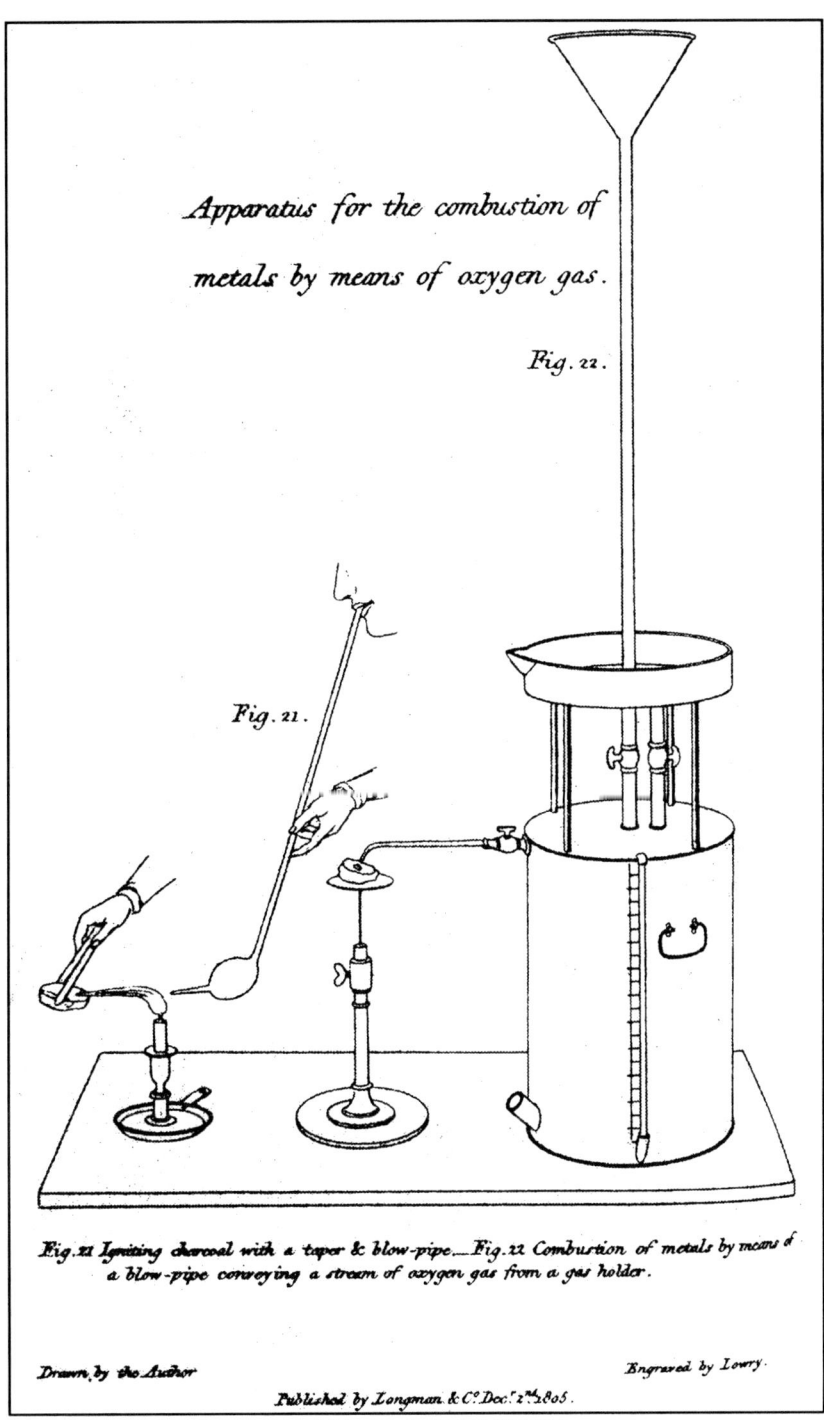

Jane Marcet's PLATE IX

79

CAROLINE
Now I think I do perfectly. Lead, for instance, will not decompose water, because it has not so strong an attraction for oxygen, as hydrogen has. Well, then, suppose the lead to be in a state of oxyd; hydrogen will take the oxygen from the lead, and unite with it to form water, because hydrogen has a stronger attraction for oxygen, than oxygen has for lead; and it is the same for all other metals which do not decompose water.

EMILY
I understand your explanation, Caroline, very well; and I imagine that it is because lead cannot decompose water that it is so much employed for pipes conveying that fluid.

MRS. B.
Certainly; lead is, on that account, particularly appropriate to such purposes; whilst, on the contrary, this metal, if it were oxydable by water, would impart to it very noxious qualities, as all oxyds of lead are more or less pernicious.

Metals and acids[1]

MRS. B.
But with regard to the oxidation of metals, there is a mode of effecting it more powerful than the former, which is by means of acids. These, you know, contain a much greater proportion of oxygen than either air or water; and will, most of them, easily yield it to metals. Have you never observed that, if you drop vinegar, lemon, or any acid on the blade of a knife, or a pair of scissors, it will immediately produce a spot of rust.

CAROLINE
Yes, often; and I am very careful to wipe off the acid immediately, to prevent the rust from forming.

EMILY
Metals have, then, three ways of obtaining oxygen; from the atmosphere; from water, and from acids.

---

[1] M1, 251-257.

MRS. B.

The first two you have already witnessed, and I shall now show you how metals take the oxygen from an acid. This bottle contains nitric acid; I shall pour some of it on this piece of copper-leaf......

CAROLINE

Oh, what a disagreeable smell!

EMILY

And what is it that produces the effervescence and that thick yellow vapour?

MRS. B.

It is the acid, which being abandoned by the greater part of its oxygen, is converted to a weaker acid, which escapes in the form of gas.

CAROLINE

And whence proceeds this heat?

MRS. B.

Indeed, Caroline, I think you might now be able to answer that question yourself.

CAROLINE

Perhaps it is that the oxygen enters into the metal in a more solid state than it existed in the acid, in consequence of which caloric is disengaged.

MRS. B.

You have found it out, you see, without much difficulty.

EMILY

The effervescence is over; therefore I suppose that the metal is now oxydated.

MRS. B.

Yes. But there is another important connection between metals and acids, with which I must make you acquainted. Metals, when in the state of oxyds, are capable of being dissolved by acids. In this operation they enter into a chemical combination with the acid, and form an entirely new compound.

CAROLINE

But what difference is there between the *oxydation* and the *dissolution* of a metal by an acid?

MRS. B.

In the first case, the metal merely combines with a portion of the oxygen taken from the acid, which is thus partly disengaged, as in the instance you have just seen; in the second case, the metal, after being previously oxydated, is actually dissolved in the acid, and enters into a chemical combination with it, without producing any further decomposition or effervescence.— This complete combination of an oxyd and an acid forms a peculiar and important class of compound salts.

EMILY

The difference between an oxyd and a compound salt, therefore, is very obvious: the one consists of a metal and oxygen; the other of an oxyd and an acid.

MRS. B.

Very well; and you will be careful to remember that the metals are incapable of entering into this combination with acids, unless they are previously oxydated: therefore, whenever you bring a metal in contact with an acid, it will first be oxydated and afterwards dissolved, provided there be a sufficient quantity of acid for both operations.

There are some metals, however, whose solution is more easily accomplished, by diluting the acid in water; and the metal will, in this case, be oxydated, not by the acid, but by the water, which it will decompose. But in proportion as the oxygen in the water oxydates the surface of the metal, the acid combines with it, washes it off, and leaves a fresh surface for the oxygen to act upon: and then other coats of oxyd are successively formed, and rapidly dissolved by the acid, which continues combining with the new-formed surfaces of oxyd until the whole of the metal is dissolved. During this process, the hydrogen gas of the water is disengaged, and flies off with effervescence.

EMILY

Was not this the manner in which sulphuric acid assisted the iron filings in decomposing water?

MRS. B.

Exactly: and it is thus that several metals, which are incapable alone of decomposing water, are enabled to do so by the assistance of an acid, which, by continually washing off the covering of oxyd, as it is formed, prepares a fresh surface of metal to act upon the water.

CAROLINE

The acid here seems to act a part not very different from that of a scrubbing brush.— But pray would not this make a good method of cleaning grates and metallic utensils?

MRS. B.

You forget that acids have the power of oxydating metals, as well as that of dissolving their oxyds; so that by cleaning a grate in this way, you would create more rust than you would destroy.

CAROLINE

True; how thoughtless I was to forget that! Let us watch the dissolution of the copper in the nitric acid; for I am very impatient to see the salt that is to result from it. The mixture is now of a beautiful blue colour; but I see no appearance of the formation of a salt; it seems to be a tedious operation.

MRS. B.

The crystallization of the salt requires some length of time to be completed; if, however, you are so impatient, I can easily show you a metallic salt already formed.

CAROLINE

But that would not satisfy my curiosity half so well as one of our own manufacturing.

MRS. B.

It is one of your own preparing that I mean to show you. When we decomposed water a few days since, by oxydation of iron filings, through the assistance of sulphuric acid, in what did the process consist?

CAROLINE

In proportion as the water yielded its oxygen to the iron, the acid combined with the new-formed oxyd, and the hydrogen escaped alone.

MRS. B.
Very well; the result, therefore, was a compound salt, formed by the combination of sulphuric acid with oxyd of iron. It still remains in the vessel in which the experiment was performed. Fetch it, and we shall examine it.

EMILY
What a variety of processes the decomposition of water, by a metal and an acid, implies! 1st, the decomposition of the water; 2ndly, the oxidation of the metal; and 3rdly, the formation of a compound salt.

CAROLINE
Here it is, Mrs. B. — What beautiful green crystals! But we do not perceive any crystals in the solution of copper in nitrous acid[1]?

MRS. B.
Because the salt is now suspended in the water which the nitrous acid contains, and will remain so until it is deposited in consequence of rest and cooling.

Crystallization[2]

EMILY
I do not understand the exact meaning of crystallization?

MRS. B.
You recollect that when a solid body is dissolved by either water or caloric, it is not decomposed; but that its integrant parts are only suspended in the solvent. When the solution is made in water, the integrant particles of the body will, on the water being evaporated, again unite into a solid mass, by force of their mutual attraction. But when the body is dissolved by caloric alone, nothing more is necessary, in order to make its particles reunite, than to reduce the temperature. And, in general, if the solvent, whether water or caloric, be slowly separated by evaporation or by

---

[1] Presumably nitric acid HNO_3 was intended, as on p. 83, rather than the weak and unstable nitrous acid, HNO_2.
[2] M1, 257-260.

cooling, and care taken that the particles be not agitated during their reunion, they will arrange themselves in regular masses, each individual substance assuming a peculiar form or arrangement; and this is what is called crystallization.

EMILY
Crystallization, therefore, is simply the reunion of the particles of a solid body that has been dissolved in a fluid.

MRS. B.
That is a very good definition of it. But I must not forget to observe, that *heat* and *water* may unite in their solvent powers; and, in this case, the crystallization may be hastened by cooling, as well as by evaporating the liquid.

CAROLINE
But if the body dissolved is of a volatile nature, will it not evaporate with the fluid?

MRS. B.
A crystallizable body, held in solution only by water, is scarcely ever so volatile as the fluid itself, and care must be taken to manage the heat, so that it may be sufficient to evaporate the water only.

I should not omit to mention that bodies, in crystallizing from their watery solution, always retain a small portion of water, which remains confined in the crystal in a solid form, and does not reappear, unless the body loses its crystalline state. This is called the *water of crystallization.*

It is also necessary that you should here more particularly remark the difference, to which we have formerly alluded, between the simple solution of bodies either in water or in caloric, and the solution of metals in acids; in the first case, the body is merely divided by the solvent into its minutest parts. In the latter, a similar effect is, indeed, produced; but it is by means of a chemical combination between the metal and the acid, in which both lose their characteristic properties. The first is a mechanical operation, the second a chemical process. We may, therefore, distinguish them by calling the first a simple solution, the other a chemical solution. Do you understand the difference?

EMILY
Yes; simple solution can affect only the attraction of aggregation. But chemical solution implies also an attraction of composition, that is to say, an actual combination between the solvent and the body dissolved.

MRS. B.
You have expressed your idea very well indeed. But you must observe, also, that whilst a body may be separated from its solution in water or caloric, simply by cooling or by evaporation; an acid can only be taken from a metal with which it is combined, by stronger affinities, which produce a decomposition.

EMILY
I think that you have rendered the difference between these two kinds of solution so obvious, that we can never confound them.

MRS. B.
Notwithstanding, however, the real difference which there appears to be between these two operations, they are frequently confounded. Indeed, several modern chemical writers, of great eminence, have even thought proper to generalize the idea of solution, and to suppress entirely the distinction introduced by the great Lavoisier, which I have taken so much pains to explain, and which I confess appears to me to render the subject much clearer.

Alloys[1]

MRS. B.
The combinations of metals with each other are called alloys; thus brass is an alloy of copper and zinc; and bronze, of copper and iron[2] &c.

EMILY
And is not pewter also a combination of metal?

MRS. B.
It is. The pewter made in this country, is mostly composed of tin, with a very small proportion of zinc and lead.

CAROLINE
Block-tin is a kind of pewter, I believe?

---

[1] M1, 263-264.
[2] In bronze, copper is alloyed with tin rather than iron.

MRS. B.

No; it is iron plated with tin, which renders it more durable, as tin will not so easily rust.

CAROLINE

Rather say *oxydate*, Mrs. B.— Rust is a word which should be exploded in chemistry.

MRS. B.

Take care, however, not to introduce the word oxydate instead of rust, in general conversation; for you will either not be understood, or you will be laughed at for your conceit.

CAROLINE

I confess that my attention is, at present, so much engrossed by chemistry, that it sometimes leads me into ridiculous observations. Every thing in nature I refer to chemistry, and have often been ridiculed for my continual allusions to it.

MRS. B.

You must be cautious and discreet in this respect, my dear, otherwise your enthusiasm, although proceeding from a sincere admiration of the science, will be attributed to pedantry.

Sympathetic ink[1]

MRS. B.

There is a beautiful green salt produced by the combination of cobalt with nitric acid, which has the singular property of forming what is called *sympathetic ink*. Characters written with this solution are invisible when cold, but when a gentle heat is applied, they assume a fine blueish-green colour[2].

---

[1] M1, 273-274.
[2] Cobalt salts are familiar today as humidity indicators, for example as monitors of the effectiveness of desiccants. The most commonly used dry form is bright ("cobalt") blue, but when it absorbs water, it turns so pale a pink as to be almost colourless. Gentle heating dries it, and restores the intense colour. Applications include museum cabinets, lab desiccators, home storage and "weather-predicting" knick-knacks, as well as agreeable lecture demonstrations based on pictures of the type suggested by Caroline.

CAROLINE

I think one might draw very curious landscapes with the assistance of this ink; I would first make a water-colour drawing of a winter scene, in which the trees should be leafless and the grass barely green; I would then trace all the verdure with the invisible ink, and whenever I chose to create spring, I should hold it before the fire, and its warmth would cover the landscape with a rich verdure.

MRS. B.

That will be a very amusing experiment, and I advise you by all means to try it.— I must now, however, take my leave of you; we have had a very long lecture, and I hope you will be able to remember it. Do not forget to write down all that you can recollect of this conversation, for the subject is of great importance, though it may not appear at first very entertaining.

CONVERSATION X

ON ALKALIES

Alkalies and earths[1]

MRS. B.
After having taken a general view of combustible bodies, we now come to the ALKALIES, and the EARTHS, which compose the class of incombustibles; that is to say, of such bodies as do not combine with oxygen at any temperature.

CAROLINE
I am afraid that incombustible substances will not be near so interesting as the others; for I have found nothing in chemistry that has pleased me so much as the theory of combustion.

MRS. B.
Do not however depreciate the incombustible bodies before you are acquainted with them; you will find that they also possess properties highly important and interesting.

Some of the earths bear so strong a resemblance in their properties to the alkalies, that it is a difficult point to know under which head to place them. The celebrated French chemist, Fourcroy[2], has classed two of them (Barytes and Strontites), with the alkalies; but, as lime and magnesia have an almost equal title to that rank, I think it better not to separate them, and therefore have adopted the common method of classing them with the earths, and of distinguishing them by the name of *alkaline earths*.

We shall first take a review of the alkalies, of which there are three species: POTASH, SODA, and AMMONIA. The first two are called *fixed alkalies*, because they exist in solid form at the temperature of the atmosphere, and require a great heat to be volatilized. The third, ammonia, has been distinguished by the name of *volatile alkali*, because its natural form is that of a gas.

---

[1] M1, 275-276.
[2] A.F.Fourcroy (1755-1809).

CAROLINE
Ammonia? I do not recollect that name in the list of simple bodies.

MRS. B.
The reason you do not find it there is that it is a compound; and if I introduce it to your acquaintance now, it is on account of its close connection with the other two alkalies, which it resembles essentially in its nature and properties. Indeed, it is not long since ammonia has resigned its place among the simple bodies, as it was not, until lately, supposed to be a compound; nor is it improbable that potash and soda may one day undergo the same fate[1], as they are strongly suspected of being compounds also.

Ash and soap[2]

MRS. B.
Wood-ashes are...... valuable for the alkali which they contain, and are used for some purposes without any further purification. Purified in a certain degree, they make what is commonly called *pearl-ash,* which is of great efficacy in taking out grease, in washing linen, &c.; for potash combines readily with oil or fat, with which it forms a compound well known to you under the name of *soap.*

CAROLINE
Really! The I should think it would be better to wash all linen with pearl-ash than with soap, as, in the latter case, the alkali, being already combined with oil, must be less efficacious in extracting grease.

MRS. B.
Its effect would be too powerful on fine linen and would injure its texture; pearl-ash is therefore only used for that which is of a strong coarse kind. For the same reason you cannot wash your hands with plain potash; but, when mixed with oil in the form of soap, it is soft as well as cleansing, and is therefore much better adapted to the purpose.

---

[1] This "fate" was not long delayed for Jane Marcet's potash (our *potassium hydroxide,* KOH) or soda (*sodium hydroxide,* NaOH). In 1807, Davy prepared the metallic elements potassium and sodium by passing an electric current through the molten alkalies.
[2] M1, 282-283.

Caustic potash, as we have already observed, acts on the skin, and animal fibre, in virtue of its attraction for water and oil, and converts all animal matter into a kind of saponaceous jelly.

Carbonat of potash[1]

EMILY
Are vegetables the only source from which potash can be derived?

MRS. B
No: for though far the most abundant in vegetables, it is by no means confined to that class of bodies, being found also on the surface of the earth mixed with various minerals, especially with earths and stones, whence it is supposed to be conveyed into vegetables by the root of the plant. It is also met with, though in very small quantities, in some animal substances. The most common state of potash is that of *carbonat*; I suppose you understand what that is?

EMILY
I believe so; though I do not recollect that you ever mentioned the word before. If I am not mistaken, it must be a compound salt formed by the union of carbonic acid with potash.

MRS. B.
Very true; you see how admirably the nomenclature of modern chemistry is adapted to assist the memory; when you hear the name of a compound, you necessarily learn what are its constituents, and when you are acquainted with the constituents, you can immediately name the compound that they form.

CAROLINE
Pray, how were bodies arranged and distinguished before this nomenclature was introduced?

MRS. B
Chemistry was then a much more difficult study; for every substance had an arbitrary name, which it derived either from the person who

---

[1] M1, 283-284.

discovered it, as *Glauber's salts*[1], for instance, or from another circumstance relative to it, though quite unconnected with its real nature, as *potash*.

These names have been retained for some of the simple bodies; for as this class is not numerous, and therefore can easily be remembered, it has not been thought necessary to change them.

EMILY

Yet I think it would have rendered the new nomenclature more complete to have methodized the names of the elementary as well as the compound bodies, though it could not have been done in the same manner. But the names of the simple substances might have indicated their nature, or, at least, some of their principle properties; and if, like the acids and compound salts, all simple bodies had a similar termination, they would have been immediately known as such. So complete and regular a nomenclature would, I think, have given a clearer and more comprehensive view of chemistry, than the present, which is a medley of old and new terms.

MRS. B.

But you are not aware of the difficulty of introducing into science an entire set of new terms; it obliges all teachers and professors to go to school again; and if some of the old names, which are least exceptionable, were not left as an introduction to the new ones, few people would have had industry and perseverance enough to submit to the study of an entirely new language; and the inferior classes of artist, who can only act from habit and routine, would, at least, for a time, have felt material inconvenience from a total change of their habitual terms. From these considerations, Lavoisier and his colleagues, who invented the new nomenclature, thought it prudent to leave a few links of the old chain, in order to connect it with the new one. Besides, you might easily conceive the inconvenience which might arise from giving a regular nomenclature to substances, the simple nature of which is always uncertain; for the new names might, perhaps, have proved to be founded in error. And, indeed, cautious as the inventors of the modern chemical language have been, it has already been found necessary to modify it in many respects. In those few cases, however, in which new names have been adopted to designate simple bodies, these names have been so contrived as to indicate one of the chief properties of the body in question; this is the case with oxygen,

---

[1] See p. 116.

which, as I explained to you, signifies the generator of acids[1]; and hydrogen, the generator of water.

Glass[2]

MRS. B.

A remarkable property of potash is the formation of glass by its fusion with silicious earth. You are not yet acquainted with this last substance further than its being in the list of simple bodies. It is sufficient, for the present, that you should know that sand and flint are chiefly composed of it; alone, it is infusible; but mixed with potash, it melts when exposed to the heat of a furnace, combines with the alkali, and runs into glass.

CAROLINE

Who would ever have supposed that the same substance that converts transparent oil into such as opake body as soap, should transform that opake body, sand, into transparent glass!

MRS. B.

The transparency, or opacity of bodies does not, I conceive, depend so much on their intimate nature, as upon the arrangement of their particles; we cannot have a more striking instance of this, than that which is afforded by the different states of carbone, which, though it commonly appears in the form of a black opake body, sometimes assumes the most dazzling transparent form in nature, that of diamond, which, you recollect, is

---

[1] The Greek preface *oxy-*, meaning *sharp*, referred not only to the taste of acids, but also to the shape of the particles which supposedly gave rise to acidic properties. Lavoisier's choice of the name *oxygène* for that part of the atmosphere which supports combustion was unfortunate because, as Davy showed, it is not present in *"muriatic acid"* (now *hydrochloric acid*, HCl). K. W. Scheele (1742-1786), who prepared the gas at about the same time, sensibly called it *fire-gas* (in German, *Feuerluft*) on account of its role in combustion, but as a humble Swedish apothecary he was slow to publish his findings. Priestley, who also prepared the gas at about that time, was a supporter of the common idea that combustion was due to the loss of an imaginary substance, of negative mass, called *phlogiston* (meaning *that which flames*). Since a flammable substance like magnesium burns better in pure oxygen than in air, Priestley argued that oxygen must mop up this phlogiston better than air does; and so oxygen must have less of its own phlogiston than air does. So he called oxygen *dephlogisticated air*. Lavoisier's inappropriate term is at least easier on tongue and ear than is Priestley's; and it was accepted by the French Academy, which arbitrated on matters intellectual, and it seems that the name *oxygen* is here to stay. See also p.33 note 2.

[2] M1, 287-289.

nothing but carbone, and which, in all probability, derives its beautiful transparency from the peculiar arrangement of its particles during their crystallization.

EMILY
I never should have supposed that the formation of glass was so simple a process as you describe it.

MRS. B.
It is by no means an easy operation to make perfect glass; for if the sand, or flint, from which the silicious earth is obtained be mixed with any metallic particles, or other substances which cannot be vitrified, the glass will be discoloured, or defaced, by opake specks.

CAROLINE
That I suppose is the reason why objects so often appear irregular and shapeless through a common glass window.

MRS. B.
This species of imperfection proceeds, I believe, from another cause. It is extremely difficult to prevent the lower part of vessels in which the materials of glass are fused, from containing a more dense vitreous matter than the upper, on account of the heavier ingredients falling to the bottom. When this happens, it occasions the appearance of veins or waves in the glass, from the difference in density of its several parts, which produces an irregular refraction of the rays of light that pass through it.

Ammonia[1]

MRS. B.
.....we must ....proceed to AMMONIA, or the VOLATILE ALKALI.

EMILY
I long to hear something of this alkali; is it not of the same nature as hartshorn?

---

[1] M1, 292-295.

MRS. B.
Yes, it is, as you will see by and by. This alkali is seldom found in nature in its pure state; it is most commonly extracted from a compound salt called *sal ammoniac*, which was formerly imported from *Ammonia*, a region of Libya, from which both the salt and the alkali derive their names. The crystals contained in this bottle are specimens of this salt, which contains a combination of ammonia and muriatic acid.

CAROLINE
Then it should be called *muriat*[1] *of ammonia*; for though I am ignorant of what muriatic acid is, yet I know that its combination with ammonia cannot but be so called; and I am surprised to see sal ammoniac inscribed on the label.

MRS. B.
That is the name by which it has been so long known, that the modern chemists have not yet succeeded in banishing it altogether; and it is still sold under that name by druggists, though by scientific chemists it is more properly called muriat of ammonia.

EMILY
By what means can ammonia be separated from the muriatic acid?

MRS. B.
By a display of chemical attractions; but this operation is complicated for you to understand, till you are better acquainted with the agency of affinities[2].

EMILY
And when it is extracted from the salt, what kind of substance is ammonia?

MRS. B.
Its natural form at the temperature of the atmosphere, when free from combination, is that of gas; and in this state it is called *ammoniacal gas*. But it mixes very readily with water, and can thus be obtained in a liquid form.

---

[1] For *muriatic acid*, see pp. 107 and 129-130.
[2] See p. 105.

CAROLINE
You said that ammonia was a compound; pray, of what principles is it composed?

MRS. B.
It was discovered a few years since by Berthollet[1], a celebrated French chemist, that it consisted of about one part of hydrogen to four parts of nitrogen. Having heated ammoniacal gas under a receiver, by causing the electrical spark to pass repeatedly through it, he found that it increased considerably in bulk, lost all its alkaline properties, and was actually converted into hydrogen and nitrogen gasses.

CAROLINE
Ammoniacal gas must, I suppose be very heavy, since it expands so much when decomposed?

MRS. B.
Compared with hydrogen gas, it certainly is; but it is considerably lighter than oxygen gas, and only about half the weight of atmospherical air. It possesses most of the properties of the fixed alkalies; but cannot be of such use in the arts on account of its volatile nature. It is, therefore, never employed in the manufacture of glass, but it forms soap with oils equally well as potash and soda: it resembles them likewise in its strong attraction for water; for which reason it can be collected in a receiver over mercury only.

---

[1] C.L.Berthollet (1748-1822).

CONVERSATION XI

ON EARTHS

Gemstones[1]

CAROLINE
I cannot conceive how such coarse materials ("earths") can be converted into such beautiful productions.

MRS. B.
We are very far from understanding all the secret resources of nature; but I do not think the spontaneous formation of crystals, which we call precious stones, one of the most difficult to comprehend..... The scarcity of many kinds of crystals, as rubies, emeralds, topazes, &c. shows that their formation is not an operation easily carried out in nature..... You know that crystallization is more regular and perfect, in proportion as the evaporation of the solvent is slow and uniform; Nature, therefore, who knows no limit of time, has, in works of this kind, an infinite advantage over any artist who attempts to imitate such productions.

EMILY
I can now conceive that the arrangement of the particles of earth, during crystallization, may be such as to occasion transparency, by admitting a free passage to the rays of light; but I cannot understand why crystallized earths should assume such beautiful colours as most of them do. Sapphire, for instance, is of celestial blue; ruby, a deep red; topaz, a brilliant yellow?

MRS. B.
Nothing is more simple than to suppose that the arrangement of their particles is such, as to transmit some coloured rays of light, and to reflect others, in which case the stone must appear the colour of the rays which it reflects. But, besides, it frequently happens that the colour of a stone is owing to the mixture of some metallic matter.

---

[1] M1, 304-305.

Silex[1]

EMILY

Pray what is the true colour of silex, which forms such a variety of different coloured substances? Sand is brown, flint is nearly black, and precious stones are of all colours?

MRS. B.

Pure silex, such as is found only in the chemist's laboratory, is perfectly white, and the various colours which it assumes, in the different substances you have just mentioned, proceed from the different ingredients with which it is mixed in them.

CAROLINE

I wonder that silex is not more valuable, since it forms the basis of so many precious stones.

MRS. B.

You must not forget that the value we set upon precious stones, depends in a great measure upon the scarcity with which nature affords them; for, were those productions either common, or perfectly imitable by art, they would no longer, notwithstanding their beauty, be so highly esteemed. But the real value of siliceous earth, in many of the most useful arts, is very extensive. Mixed with clay, it forms the basis of all the various kinds of earthen ware, from the most common utensils to the most refined ornaments.

EMILY

And we must not forget its importance in the formation of glass with potash.

MRS. B.

Nor should we omit to mention, likewise, many other important uses of silex, such as being the chief ingredient of some of the most durable cements, of mortar, &c.

---

[1] M1, 311-312.

Alumine[1]

MRS. B.
We will now hasten to proceed to the other earths, for I am rather apprehensive of your growing weary of this part of the subject.

CAROLINE
The history of the earths is not quite so entertaining as that of the other simple substances.

MRS. B.
Perhaps not; but it is absolutely indispensable that you should know something of them; for they form the basis of so many interesting and important compounds, that their total omission would throw great obscurity on our general outline of chemical science. We shall, however, review them in as cursory a manner as the subject can admit of.

ALUMINE derives its name from a compound salt called *alum*, of which it forms the basis.

CAROLINE
But it ought to be just the contrary, Mrs. B. — ; the simple body should give, instead of taking, its name from the compound.

MRS. B.
Very true, my dear; but as the compound salt was known long before its base was discovered, it was natural when that earth was at length separated from the acid, that it should derive its name from the compound from which it was obtained. However, to remove your scruples, we will call the salt according to the new nomenclature, *sulphat of alumine*. From this combination, alumine may be obtained in its pure state; it is then soft to the touch, makes a paste with water, and hardens in the fire. In nature, it is found chiefly in clay, which contains a very considerable proportion of this earth; and it is very abundant in fuller's earth, slate, and a variety of other mineral productions. There is indeed scarcely any mineral substance more useful to mankind than alumine. In the state of clay, it forms large strata of the earth, gives consistency to the soil of valleys....

---

[1] M1, 313-315.

The solid compact soils, such as are fit for corn, owe their consistence in a great measure to alumine; this earth is therefore used to improve sandy or chalky soils, which do not retain a sufficient quantity of water for vegetation.

Alumine is the most essential ingredient in all potteries. It enters into the composition as brick, as well as in that of the finest china; the addition of silex and water hardens it, renders it susceptible of a degree of vitrification, and makes it perfectly fit for its various purposes.

CAROLINE

I can scarcely conceive that brick and china should be made of the same materials.

MRS. B.

Brick consists almost entirely of baked clay; but a certain proportion of silex is essential to the formation of earthen or stone ware. In common potteries, sand is used for that purpose; a more pure silex is, I believe, necessary for the composition of porcelain, as well as a finer kind of clay; and these materials are, no doubt, more carefully prepared, and curiously wrought, in the one case than in the other. Porcelain owes its beautiful semi-transparency to a commencement of vitrification.

Lime-water and carbonic acid gas[1]

MRS. B.

I said that the attraction of lime for carbonic acid was so strong, that it would absorb it from the atmosphere. We may see this effect by exposing a glass of lime-water to the air; the lime will then separate from the water, combine with the carbonic acid, and reappear on the surface in the form of a white film, which is carbonat of lime, commonly called *chalk*.

CAROLINE

Chalk is, then, a compound salt! I never should have supposed that those immense beds of chalk, that we see in many parts of the country, were a salt.— Now, the white film begins to appear on the surface of the water; but it is far from resembling hard solid chalk.

---

[1] M1, 320-323.

MRS. B.
That is owing to its state of extreme division; in a little time it will collect into a more compact mass, and subside at the bottom of the glass.

If you breathe into lime-water, the carbonic acid, which is mixed with the air you expire, will produce the same effect. It is an experiment very easily made— I shall pour some lime-water into this glass tube, and, by breathing repeatedly into it, you will soon perceive a precipitation of chalk—

EMILY
I see already a small white cloud formed.

MRS. B.
It is composed of minute particles of chalk; at present it floats in the water, but it will soon subside.

Carbonat of lime, or chalk, you see, is insoluble in water since the lime which was dissolved reappears when converted into chalk; but you must take notice of a very singular circumstance, which is, that chalk is soluble in water impregnated with carbonic acid.

CAROLINE
It is very curious, indeed, that carbonic acid gas should render lime soluble in one instance, and insoluble in the other!

MRS. B.
I have here a bottle of Seltzer water, which, you know, is strongly impregnated with carbonic acid— let us pour a little of it into a glass of lime-water.— You see that it immediately forms a precipitate of carbonat of lime?

EMILY
Yes, a white cloud appears.

MRS. B.
I shall now pour an additional quantity of the Seltzer water into the lime-water—

EMILY
How singular! The cloud is redissolved, and the liquid is again transparent.

MRS. B.

All the mystery depends on this circumstance, that carbonat of lime is soluble in carbonic acid, whilst it is insoluble in water; the first quantity of carbonic acid, therefore, which I introduced into the lime-water, was employed in forming the carbonat of lime, which remained visible, until an additional quantity of carbonic acid dissolved it. Thus, you see, when lime and carbonic acid are in the proper proportions to form chalk, the white cloud appears, but when the acid predominates, the chalk is no sooner formed than it is dissolved.

CAROLINE

That is now the case; but let us try whether a further addition of lime-water will again precipitate the chalk.

EMILY

It does indeed! The cloud reappears, because, I suppose, there is now no more of the carbonic acid than is necessary to form chalk; and, in order to dissolve the chalk, a superabundance of acid is required.

MRS. B.

We have, I think, carried this experiment far enough; every repetition would but exhibit the same appearances.

Selections from

VOL. II

ON COMPOUND BODIES

CONVERSATION XII

ON THE ATTRACTION OF COMPOSITION

The Laws of Chemical Attraction[1]

MRS. B.
Having completed our examination of the simple or elementary bodies, we are now to proceed to those of a compound nature; but before we enter on this extensive subject, it will be necessary to make you acquainted with the laws by which chemical combinations are governed.

You recollect, I hope, what we have formerly said of the nature of the attraction of composition, or chemical attraction, or affinity, as it is also called?

EMILY
Yes, I think perfectly; it is the attraction that subsists between bodies of a different nature, which occasions them to combine and form a compound, when they come into contact.

MRS. B.
Very well; your definition comprehends the 1st law of chemical attraction, which is, that *it takes place only between bodies of a different nature*; as, for instance, between an acid and an alkali; between oxygen and a metal, &c. ......

...The 2d law of chemical attraction is, that *it takes place only between the minute particles of bodies*; therefore, the more you divide the particles of the bodies to be combined, the more readily they act on each other.

CAROLINE
... It was for this purpose, you said, that you used iron filings, in preference to wires or pieces of iron, for the decomposition of water..........

---

[1] M2, 1-10.

MRS. B.

.....The 3d law of chemical combination *is, that it can take place between two, three, four, or even a greater number of bodies.....* You will soon become acquainted with a great variety of these complicated compounds. The 4th law of chemical attraction is, that *a change of temperature always takes place at the moment of combination........*

I am going to show you a very striking instance of the change of temperature arising from the combination of different bodies.— I shall pour some nitrous acid[1] on this small quantity of oil of turpentine— the oil will instantly combine with the oxygen of the acid, and produce a considerable change of temperature.

CAROLINE

What a blaze! The temperature of the oil and the acid must be elevated indeed to produce such a violent combustion........

MRS. B.

The 5th law of chemical attraction is that *the properties which characterize bodies, when separate, are altered or destroyed by their combination.*

CAROLINE

Certainly; what, for instance, can be so different from water than oxygen and hydrogen gasses?

EMILY

Or what more unlike sulphat of iron than iron or sulphuric acid?

CAROLINE

But of all metamorphoses, that of sand and potash into glass, is the most striking!

MRS. B.

Every chemical combination is an illustration of this rule. But let us proceed—
The 6th law is, that *the force of chemical affinity, between the constituents of a body, is estimated by that which is required for their separation.* ...

---

[1] Jane Marcet must have used nit*ric* acid for this demonstration; but readers are advised **NOT** to test this for themselves.

EMILY

But, Mrs. B., you speak of estimating the force of attraction between bodies, by the force required to separate them; how can you measure these forces?

MRS. B.

They cannot be precisely *measured,* but they are comparatively ascertained by experiment, and can be represented by numbers which express the relative degrees of attraction.

The 7th law is, that *bodies have amongst themselves different degrees of attraction.* Upon this law (which you may have discovered for yourselves long since), the whole science of chemistry depends; for it is by means of the various degrees of affinity which bodies have for each other, that all the chemical compositions and decompositions are effected. ....Whenever the decomposition of a body is effected by a single new substance, it is said to be effected by simple elective attractions[1]. But it often happens that no simple substance will decompose a body, and that, in order to effect this, you must offer to the compound body which is itself composed of two, or sometimes three principles, which would not, each separately, perform the decomposition. In this case there are two new compounds formed in consequence of a reciprocal decomposition and recomposition. All instances of this kind are called *double elective attractions.*

CAROLINE

I confess I do not understand this clearly[2].

---

[1] Elective affinity was a concept which attempted to quantify the tendency of a substance to enter into chemical reaction. In late eighteenth century France, numerical values of affinity were assigned to many substances in the (unfulfilled) hope of revealing a general pattern of chemical reactivity.

[2] It appeared that Caroline's worries were soon dispelled by Mrs. B.'s subsequent explanation, which is not included here; but a modern reader, with access to the first edition (pp. M2, 10-13), might not be so easily convinced.

CONVERSATION XIII

ON COMPOUND BODIES

The classification of acids[1]

CAROLINE
The modern nomenclature[2] must be of immense advantage in pointing out so easily the nature of the acids, and their various degrees of oxidation.

MRS. B.
Certainly. But great as are the advantages of the new nomenclature in this respect, it is not possible to apply it in its full extent to all acids, because the radicals or bases of some of them are still unknown.

CAROLINE
If you are acquainted with the acid, I cannot understand how its basis can remain unknown; you have only to separate the oxygen from it by elective attractions, and the basis must remain alone?

MRS. B.
This is not always so easily accomplished as you imagine; for there are some acids which no chemist has hitherto been able to decompose by any means whatsoever. It appears that the bases of these undecompounded acids have so strong an attraction for oxygen, that they will yield it to no other substance; and in that case, you know, the efforts of the chemist are vain.

EMILY
But if these acids have never been decomposed, should they not be classed with the simple bodies; for you have repeatedly told us that the simple bodies are rather such as the chemists have been unable to decompose, than such as are really supposed to consist of only one principle?

---
[1] M2, 17-19.
[2] See pp. 58-59.

MRS. B.

Analogy affords us so strong a proof of the compound nature of the undecompounded acids, that I could never reconcile myself to classing them with the simple bodies, though this division has been adopted by several chemical writers. It is certainly the most strictly regular; but, as a systematical arrangement is of use only to assist the memory in retaining facts, we may, I think, be allowed to deviate from it when there is danger of producing confusion by following it too closely; and this, I believe, would be the case, if you were taught to consider undecompounded acids as elementary bodies.

EMILY

I am sure you would not deviate from the methodical arrangement without good reason.— But pray what are the names of these undecompounded acids?

MRS. B.

There are three of that description:
  The *muriatic acid.*
  The *boracic acid.*
  The *fluoric acid.*

Since these acids[1] cannot derive their names from their radicals, they are called after the compound substances from which they are extracted.

CAROLINE

We have heard of a great variety of acids; pray how many are there in all?

MRS. B.

I believe there are reckoned at present thirty-four, and their number is constantly increasing, as the science improves; but the most important, and those to which we shall almost entirely confine our attention, are but few. I shall, however, give you a general view of the whole; and then we shall more particularly examine those that are the most essential.

---

[1] Muriatic acid (now called *hydrochloric acid* or aqueous *hydrogen chloride*) was obtained from sea salt (*sodium chloride*) and named after the Latin, *muria*, for brine. Like fluoric acid (*hydrofluoric acid* or aqueous *hydrogen fluoride*) it was later found to contain no oxygen (see pp. 128, 148).

This class of bodies was formerly divided into mineral, vegetable and animal acids, according to the substances from which they were extracted.

CAROLINE

That I should think must have been an excellent arrangement; why was it altered?

MRS. B.

Because in many cases it produced confusion. In which class, for instance, would you place carbonic acid?

CAROLINE

Now I see the difficulty. I would be at a loss where to place it, as you have told us that it exists in the animal, vegetable, and mineral kingdoms.

EMILY

There would be the same objection with respect to phosphoric acid, which, though obtained chiefly from bones, can also, you said, be found in small quantities in stones, and likewise in some plants.

MRS. B

You, see therefore, the propriety of changing this mode of classification. These objections to not exist in the present nomenclature; for the composition and nature of each individual acid is in some degree pointed out, instead of the class of bodies from which it is extracted; and with regard to the more general division of acids, they are classed under these four heads:

1st, Acids of known and simple bases, which are formed by the union of these bases with oxygen. They are the following:

The *Sulphuric*
     Carbonic
     Nitric
     Phosphoric
     Arsenical
     Tungstenic
     Molybdenic

2ndly, Those of unknown bases:

> The *Muriatic*
> *Boracic*
> *Fluoric*

These two classes comprehend the most anciently known, and most important acids. The sulphuric, nitric, and muriatic, were formerly, and are still frequently, called *mineral* acids.

3rdly, Acids that have double or binary radicals, and which consequently consist of triple combinations, whose common radical[1] is a compound of hydrogen and carbone.....

The 4th class of acids consists of those which have triple radicals and are therefore of a still more compound nature. This class comprehends the animal acids, which are

> Lactic[2],
> Prussic[3],
> Formic[4].

---

[1] Mrs. B. explains that the acids differ in the ratio of hydrogen to carbone in the radical. Many of her examples are derived from vegetables, and some are familiar to us by trivial names which are still in use, e.g. *acetic, oxalic, citric, benzoic*......

[2] But lactic acid (from milk), like the vegetable acids, contains only carbon and hydrogen in addition to the oxygen which was then supposed to be prerequisite for acidity.

[3] The potassium salt of *prussic* (now called *hydrocyanic*) acid was made by heating potassium hydroxide with blood. But since the acid (HCN) can also be made directly by heating ammonia (NH_3) with carbon, it cannot contain any oxygen. The acid is related to the pigment *Prussian Blue*, which was first made in Berlin, the capital of what was then Prussia.

[4] Formic, was first made from ants and, like lactic acid, contains only hydrogen, carbon and oxygen. But animal origin does not necessarily confer complexity; and we now know this acid as *methanoic*, acid (H.COOH).

CONVERSATION XIV

ON THE COMBINATIONS OF OXYGEN WITH SULPHUR AND WITH PHOSPHORUS; AND OF THE SULPHATS AND PHOSPHATS

Concentrated sulphuric acid[1]

MRS. B.
In addition to the general survey which we have taken of acids, I think you will find it interesting to examine individually a few of the most interesting of them, and likewise some of their principal combinations with the alkalies, alkaline earths and metals. The first of these acids, in point of importance, is the SULPHURIC, formerly called *oil of vitriol*.

CAROLINE
I have known it a long time by that name, but had no idea that it was the same fluid as sulphuric acid. What resemblance or connection can there be between oil of vitriol and this acid?

MRS. B.
Vitriol is the common name for sulphat of iron, a salt which is formed by the combination of sulphuric acid and iron; the sulphuric acid was formerly obtained by distillation from this salt, and it very naturally received its name from the substance which afforded it.

CAROLINE
But is it still usually called oil of vitriol?

MRS. B.
Yes; a sufficient length of time has not yet elapsed, since the invention of the new nomenclature, for it to be generally disseminated; but, as it is adopted by all scientific chemists, there is every reason to suppose that it will gradually become universal. When I received this bottle from the

---

[1] M2, 28-33.

chemist's, the name written on the label was *oil of vitriol*; but, as I knew you were very punctilious in regard to nomenclature, I changed it, and substituted the modern name.

EMILY

This acid has neither colour nor smell, but it appears much thicker than water.

MRS. B.

It is twice as heavy as water, and has, you see, the appearance and consistence of oil.

CAROLINE

And it is probably from this circumstance that it has been called an oil, for it can have no real claim to that name, as it does not contain either hydrogen or carbone, which are the essential constituents of oil[1].

MRS. B.

Certainly; and therefore it would be the more absurd to retain a name which owed its origin to such a mistaken analogy.

Sulphuric acid, in its purest state, would be a concrete substance, but its attraction for water is such that it is impossible to preserve it in that state; it is, therefore, always seen in a liquid form, such as you here find it. One of the most striking properties of sulphuric acid is that of evolving a considerable quantity of heat when mixed with water; this I have already shown you.

EMILY

Yes, I recollect it; but what was the degree of heat produced by that mixture?

MRS. B.

The thermometer may be raised by it to 300°, which is considerably above the degree of boiling water[2].

CAROLINE

Then water might be made to boil in that mixture?

---

[1] See p.74, note 1.
[2] As Jane Marcet uses the Fahrenheit scale, her 300° is about 149°C. At unit atmospheric pressure, water boils at 212°F (100°C).

MRS. B.

Nothing more easy, provided that you employ sufficient quantities of acid and of water, and in the due proportions. The greatest heat is produced by a mixture of one part of water to four of acid: we shall make[1] a mixture of these proportions, and immerse this thin glass tube, which is full of water, into it.

CAROLINE

The vessel feels extremely hot, but the water does not boil yet.

MRS. B.

You must allow some time for the heat to penetrate the tube, and raise the temperature of the water to the boiling point—

CAROLINE

Now it boils— and with increasing violence.

MRS. B.

But it will not continue boiling long; for the mixture gives out heat only while the particles of the water and the acid are mutually penetrating each other: as soon as the new arrangement of these particles is effected, the mixture will gradually cool and the water return to its former temperature.

You have seen the manner in which sulphuric acid decomposes all combustible substances, whether animal, vegetable, or mineral, and burns them by means of its oxygen?

CAROLINE

I have very unintentionally repeated the experiment on my gown, by letting a drop of acid fall upon it, and it has made a terrible stain, which, I suppose, will never wash out.

MRS. B.

No, indeed; for before you can put it into water, the spot will become a hole, as the acid has literally burnt the muslin.

---

[1] It is strange that Jane Marcet does not stress the importance of adding the **ACID TO WATER**, since the converse would generate enough heat to make the mixture bubble and probably spurt concentrated acid into the user's face.

CAROLINE
So it has, indeed! Well, I will fasten the stopper and put the bottle away, for it is a dangerous substance.— Oh, now I have done worse still, for I have spilt some on my hand!

MRS. B
It is then burned, as well as your gown, for you know that oxygen destroys animal as well as vegetable matter; and, as far as the decomposition of the skin of your finger is effected, there is no remedy; but, by washing it immediately in water, you will dilute the acid, and prevent any further injury.

CAROLINE
It feels extremely hot, I assure you.

MRS. B.
You have now learned, by experience, how cautiously this acid must be used. You will soon become acquainted with another acid, the nitric, which although it produces less heat on the skin, destroys it still quicker, and make upon it an indelible stain. You should never handle substances of this kind, without previously dipping your fingers in water, which will weaken their caustic effects. But since you will not expose your fingers again, I must put in the stopper, for the acid attracts the moisture from the atmosphere, which would destroy its strength and purity.

Antidotes to acid poisoning[1]

MRS. B.
From its powerful properties, and from the various combinations into which it enters, sulphuric acid is of great importance in many of the arts.

It is also used in medicine in a state of great dilution; for were it taken internally, in a concentrated state, it would prove a most dangerous poison.

CAROLINE
I am sure it would burn the throat and stomach.

---

[1] M2, 34-35.

MRS. B.
Can you think of any thing that would prove an antidote to this poison?

CAROLINE
A large draught of water to dilute it.

MRS. B.
That would certainly weaken the power of the acid, but it would increase the heat to an intolerable degree. Do you recollect nothing that would destroy its deleterious effects more effectually?

EMILY
An alkali might, by combining with it; but then, a pure alkali is itself a poison, on account of its causticity.

MRS. B.
There is no necessity that the alkali should be caustic. Soap, in which it is combined with oil: or magnesia, either in the state of carbonat, or mixed with water, would prove the best antidotes[1].

Sulphurous acid as a bleaching agent[2]

MRS. B.
Sulphurous acid is readily absorbed by water: and in this liquid state it is found particularly useful in bleaching linen and woollen cloths, and is much used in manufactures for these purposes. I can show you its effect in destroying colours, by taking any iron-mould, or vegetable stain— I think I see a spot on your gown, Emily, on which we may try the experiment.

---

[1] This passage was said to have saved the life of a young sister of the educationalist Maria Edgeworth. The girl had swallowed an apparently life-threatening quantity of some acid, but medical help was of no avail. Then someone remembered that, in *Conversations on Chemistry*, magnesia had been recommended as an antidote; and so the girl was dosed with this, and she recovered. Maria Edgeworth wrote to thank the anonymous author, and the two women later met in London and became friends. See Armstrong, E. V., *J. Chem. Educ.*, **15**, 53 (1938).
[2] M2, 37-38.

EMILY
It is the stain of mulberries; but I shall be almost afraid of exposing my gown to the experiment, after seeing the effect which the sulphuric acid produced on that of Caroline—

MRS. B.
There is no danger from the sulphurous; but the experiment must be made with great caution, for, during the formation of sulphurous acid by combustion, there is always some sulphuric acid produced.

CAROLINE
But where is your sulphurous acid?

MRS. B.
We may easily prepare some ourselves, simply by burning a match[1]; we must first wet the stain with a little water, and now hold it in this way, at a little distance, over the lighted match: the vapour that arises from it is sulphurous acid, and the stain, you see, gradually disappears.

EMILY
I have frequently taken out stains by this means, without understanding the nature of the process. But why is it necessary to wet the stain before exposing it to the acid fumes?

MRS. B.
The moisture attracts and absorbs the sulphurous acid: and it serves likewise to dilute any particles of sulphuric acid which might injure the linen.

Sulphats of potash and soda[2]

MRS. B.
The most important of the salts, formed by the combinations of sulphuric acid, are, 1st, *sulphat of potash*, formerly called *sal polycrest*; this

---

[1] These were probably strips of wood tipped with sulphur, which were used as tapers. They were usually ignited by touching them with a hot poker or ember. (As early as 1680, Robert Boyle had found that a sulphur match could be ignited by pulling it between two folds of coarse paper which had been treated with white phosphorus, but it was at least a century before the process was developed for commercial use.) See p.64.

[2] M2, 39-41.

is a very bitter salt, much used in medicine; it is found in the ashes of most vegetables, but it may be prepared artificially by the immediate combination of sulphuric acid with potash. This salt is easily soluble in boiling water. Solubility is, indeed, a property common to all salts; and they always produce cold on melting.[1]

EMILY
That must be owing to the caloric which they absorb in passing from a solid to a fluid form.

MRS. B.
That is, at least, the most probable explanation.

Sulphat of soda, commonly called Glauber's salt[2], is another medicinal salt, which is still more bitter than the preceding. We must prepare some of these compounds, that you may observe the phenomena which takes[3] place during their formation. We need only pour some sulphuric acid over the soda which I put into this glass.

CAROLINE
What an amazing heat is disengaged.– I thought you said that cold was produced by the melting[4] of salts.

MRS. B.
But you must observe that we are now *making*, not *melting* a salt. Heat is disengaged during the formation of compound salts, because the acid goes into a more dense state in the salt than that in which it existed before.

---

[1] Nowadays, chemists restrict the term *melting* to liquefaction by heat in the absence of an added solvent, and in the context above would replace it by the word *dissolution*.
[2] Named after the German chemist J. R. Glauber (1604-1670).
[3] *Sic*. But probably many of her readers also lacked a classical education.
[4] See p.17.

CONVERSATION XV

OF THE NITRIC AND CARBONIC ACIDS: OR THE COMBINATION OF OXYGEN WITH NITROGEN AND CARBONE; AND OF THE NITRATS AND CARBONATS

Nitric acid[1]

MRS. B.
Nitric acid has been used in the arts from time immemorial, but it is not more than twenty-five years that its chemical nature has been ascertained. The celebrated Mr. Cavendish[2] discovered that it consisted of 20 parts of nitrogen and 80 of oxygen. These principles, in their gaseous state, combine at a high temperature; and this may be effected by repeatedly passing the electric spark through a mixture of the two gasses.

EMILY
The nitrogen and oxygen gasses, that compose the atmosphere, do not combine, I suppose, because their temperature is not sufficiently elevated?

CAROLINE
But in a thunderstorm, when lightning repeatedly passes through them, may it not produce nitric acid? We should be in a strange situation if a violent storm should at once convert the atmosphere to nitric acid.

MRS. B.
There is no danger of it, my dear; the lightning can affect but a very small portion of the atmosphere, and though it is supposed that it does occasionally produce a little nitric acid, yet this never happens to such an extent as to be perceivable.

---

[1] M2, 51-53.
[2] Cavendish (see p.46) prepared nitric acid in 1784, but his figures show that the nitrogen was not completely oxidised.

EMILY
But how could the nitric acid be known, and used, before the method of combining its constituents was discovered?

MRS. B.
Before that period the nitric acid was obtained (and it is indeed still extracted, for the common purposes of art) from the compound salt which it forms with potash, which is commonly called *nitre*.

CAROLINE
Why is it so called? Pray, Mrs. B., let these old unmeaning names be entirely given up, by us at least; and let us call this salt *nitrat of potash*.

MRS. B.
With all my heart; but it is necessary that I should, at least, mention the old names, and more especially those that are yet in common use; otherwise, when you meet with them, you would not be able to understand their meaning.

EMILY
And how is the acid obtained from this salt?

MRS. B.
By the intervention of sulphuric acid, which combines with the potash, and sets the nitric acid at liberty. This I can easily show you, by mixing some nitrat of potash and sulphuric acid in this retort, and heating it over a lamp; the nitric acid will come over in the form of vapour, which we shall collect in a glass bell. This acid is commonly called *aqua fortis*, if Caroline will allow me to mention that name.

CAROLINE
I have often heard that aqua fortis will dissolve almost all metals; it is no doubt because it yields its oxygen so easily.

MRS. B.
Yes; and from this powerful solvent property, it derived the name of aqua fortis, or strong water.

Aurora borealis[1]

MRS. B.
Mr. Lybe[2] a late French writer on natural philosophy, has supposed that the aurora borealis may be occasioned by the red fumes of nitrous acid, formed by the combination of the oxygen and nitrogen gasses in the atmosphere, when their temperature is raised by lightning.

CAROLINE
That appears to me a very natural conjecture; for the streaks of the aurora borealis have very much the colour and appearance of the fumes of nitrous acid.

EMILY
But if this were the case, why should these fumes be confined to the vicinity of the poles?

MRS. B.
Lybe accounts for that circumstance by supposing, that in warm climates, at a distance from the poles, a considerable quantity of hydrogen gas is exhaled by the decomposition of animal and vegetable matter, a decomposition which cannot take place at the very low temperature of the polar regions; and that this hydrogen gas, floating in the upper regions of the atmosphere, is, by the occasional effect of the electric fluid, made able to unite with the oxygen, which is thus prevented from combining with the nitrogen, for which it has less affinity. In warm climates, therefore, the effect of lightning is to produce rain instead of aurora borealis.

EMILY
I like this theory very much, as it accounts likewise for the greater quantity of rain that falls in warm climates.

MRS. B.
It is extremely ingenious, but we are at present too little acquainted with the nature of meteorological phenomena to be able to place much reliance on it.

---

[1] M2, 55-56.
[2] No trace can be found of the imaginative Mr. Lybe.

"Exhilarating Gas"[1]

MRS. B.

It remains for me only to mention another curious modification of oxygenated nitrogen, which has been distinguished by the name of *gaseous oxyd of nitrogen*. It is but lately that this gas[2] has been accurately examined, and its properties have been investigated chiefly by Mr. Davy. It has obtained also the name *exhilarating gas*, from the very singular property, which that gentleman discovered in it, of elevating the animal spirits, when inhaled into the lungs, to a degree sometimes resembling delirium or intoxication.

CAROLINE

It is respirable, then?

MRS. B.

It can scarcely be called respirable, as it would not support life for any length of time; but it may be breathed for a few moments without any other effects, than the singular exhilaration of spirits I have just mentioned. It affects different people, however, in a very different manner. Some become violent, even outrageous; others experience a languor, attended with faintness[3]; but most agree in opinion, that the sensations it excites are extremely pleasant.

CAROLINE

I think I should like to try it— how do you breathe it?

MRS. B.

By collecting the gas in a bladder, to which a short tube with a stop-cock is adapted; this is applied to the mouth with one hand, whilst the nostrils are kept closed with the other, that the common air may have no access. You then alternately inspire, and expire the gas, till you perceive its effects. But I cannot consent to your making the experiment; for the nerves

---

[1] M2, 61-65.

[2] Subsequently called *dinitrogen monoxide* or *nitrous oxide*, N_2O. It was first prepared in 1772 by Joseph Priestley, who reported that it supported combustion (because, as we now know, it breaks done into dioxygen and dinitrogen when heated).

[3] Although Davy reported that it could produce anaesthesia, it was at first used only for entertainment. Its anaesthetic properties were rediscovered at Harvard in the early 1840s, and then were widely used in surgery. A century later, nitrous oxide was still being used as a light anaesthetic for dentistry; and even today it serves as an agreeable sedative and mild analgesic in childbirth and in children's dentistry.

are sometimes unpleasantly affected by it, and I would not run any risk of that kind.

EMILY

I should like, at least, to see somebody breathe it; but pray by what means is this curious gas obtained?

MRS. B.

It is procured from *nitrat of ammonia*, an artificial salt which yields this gas on the application of a gentle heat.— I have put some of the salt into a retort, and by the aid of a lamp the gas will be extricated—

CAROLINE

Bubbles of air begin to escape through the neck of the retort into the water apparatus; will you not collect them?

MRS. B.

The gas that first comes over is never preserved, as it consists of little more then the common air which was in the retort; besides, there is always in this experiment a quantity of watery vapour which must come away before the nitrous oxyd appears.

EMILY

Watery vapour! Whence does that proceed? There is no water in nitrat of ammonia?

MRS. B.

You must recollect that there is in every salt a quantity of water of crystallization, which may be evaporated by heat alone. But, besides this, water is actually generated in this experiment, as you will see presently. But first tell me, what are the constituent parts of nitrat of ammonia?

EMILY

Ammonia, and nitric acid: this salt, therefore, contains three different elements, nitrogen and hydrogen, which produces the ammonia; and oxygen, which, with nitrogen, forms the acid.

MRS. B.

Well, then, in this process the ammonia is decomposed; the hydrogen quits the nitrogen to combine with some of the oxygen of the nitric acid, and forms with it the watery vapour which is now coming over. When that is effected, what will you expect to find?

EMILY
Nitrous acid instead of nitric acid, and nitrogen instead of ammonia.

MRS. B.
Exactly so; and the nitrous acid and the nitrogen combine, and form the gaseous oxyd of nitrogen, in which the proportion of oxygen is 37 parts to 63 of nitrogen.

You may have observed, that for a little while no bubbles of air have come over, and we have perceived only a stream of vapour condensing as it issued into the water.— Now bubbles of air again make their appearance, and I imagine that by this time all the watery vapour is come away, and that we may begin to collect the gas. We may try whether it is pure by filling a phial with it, and plunging a taper into it— yes, it will do now; for the taper burns brighter than in common air, and with a greenish flame.

CAROLINE
But how is that? I thought no gas would support combustion but oxygen?

MRS. B.
Or any gas that contains oxygen, and is ready to yield it, which is the case with this to a considerable degree; it is not, therefore, surprising that it should accelerate the combustion of a taper.

You can see that the gas is now produced in great abundance; we shall collect a large quantity of it, and I dare say that we shall find some of the family who will be curious to make the experiment of respiring it.

Gunpowder[1]

MRS. B.
Whilst this process [*the collection of gaseous oxyd of nitrogen*] is going on, we may take a general survey of the most important combinations of nitric and nitrous acids with the alkalies.

The first of these is *nitrat of potash*, commonly known as *nitre* or *saltpetre*.

---

[1] M2, 65-67.

CAROLINE
Is that not the salt with which gunpowder is made?

MRS. B.
Yes. Gunpowder is a mixture of five parts of nitre to one of sulphur, and one of charcoal.— Nitre from its great proportion of oxygen, and from the facility with which it yields it, is the basis of most detonating compositions.

EMILY
But what is the cause of the violent detonation of gunpowder when set fire to?

MRS. B.
Detonation may proceed from two causes; the sudden formation or destruction of an elastic fluid. In the first case, when either a solid or a liquid is instantaneously converted into an elastic fluid, the prodigious and sudden expansion of the body strikes the air with great violence, and this concussion produces the sound called detonation.

CAROLINE
That I comprehend very well; but how can a similar effect be produced by the destruction of a gas?

MRS. B.
A gas can be destroyed only by condensing it to a liquid or solid state; when this takes place suddenly, the gas, in assuming a new and more compact form, produces a vacuum into which the surrounding air rushes with great impetuosity; and it is by that rapid and violent motion that the sound is produced. In all detonations, therefore, gasses are either suddenly formed, or destroyed. In that of gunpowder, can you tell me which of these two circumstances take place?

EMILY
As gunpowder is a solid, it must, of course, produce the gasses in its detonation; but how, I cannot tell.

MRS. B.

The constituents of gunpowder, when heated to a certain degree, enter into a number of new combinations, and are instantly converted into a variety of gasses, the sudden expansion of which gives rise to the detonation.

CAROLINE

And in what instance does the destruction or condensation of gasses produce detonation?

MRS. B.

I can give you one with which you are well acquainted; the sudden combination of the oxygen and hydrogen gasses.

CAROLINE

True; I recollect perfectly that hydrogen detonates with oxygen when the two gasses are converted into water.

Carbonic acid gas[1]

MRS. B.

Carbonic acid gas is found very abundantly in nature; it is supposed to form about one hundredth part of the atmosphere, and is constantly produced by the respiration of animals...... It is contained in a state of great purity in certain caves, such as the *Grotto del Cane*, near Naples.

EMILY

I recollect having read an account of that grotto, and the cruel experiments made on the poor dogs, to gratify the curiosity of strangers. But I understood that the vapour exhaled by this cave was called *fixed air*.

MRS. B.

That is the name by which carbonic acid was known before its chemical composition was discovered.–This gas is more destructive of life than any other; and if the poor animals that are submitted to its effects, are not plunged into cold water as soon as they become senseless, they do not recover. It extinguishes flame instantaneously. I have collected some in this glass, which I will pour over the candle.

---

[1] M2, 72-75.

CAROLINE

This is extremely singular—it seems to extinguish as if by enchantment, as the gas is invisible. I should never have imagined that a gas could have been poured like a liquid.

MRS. B.

It can be done with carbonic acid only, as no other gas is sufficiently heavy to be susceptible of being poured out in atmospherical air without mixing with it.

EMILY

Pray by what means did you obtain this gas?

MRS. B.

I procured it from marble. Carbonic acid gas[1] has so strong an attraction for all the alkalies and alkaline earths, that these are always found in nature in the state of carbonats. Combined with lime, this acid forms chalk, which may be considered as the basis of all kinds of marbles, and calcareous stones. From these substances carbonic acid is easily separated, as it adheres so slightly to its combinations, that the carbonats are all decomposable by any of the other acids. I can easily show you how I obtained this gas; I poured some diluted sulphuric acid over pulverized marble in this bottle (the same which we used the other day to make hydrogen gas), and the gas escaped through the tube connected with it; the operation still continues, as you may easily perceive—

EMILY

Yes, it does; there is a great fermentation in the glass vessel. What singular commotion is excited by the sulphuric acid taking possession of the lime, and driving out the carbonic acid!

CAROLINE

But did the carbonic acid exist in a gaseous state in the marble?

MRS. B.

Of course not; the acid, when in a state of combination, is capable of existing in solid form.

---

[1] Carbon dioxide, CO_2.

CAROLINE
Whence, then, does it obtain the caloric necessary to convert it to a gas?

MRS. B.
It may be supplied in this case by the mixture of sulphuric acid and water, which produces an evolution of heat, even greater than is required for the purpose; since, as you may perceive by touching the glass vessel, a considerable quantity of the caloric disengaged becomes sensible. But a supply of caloric may be obtained also from a diminution of capacity for heat, occasioned by the new combination which takes place; and, indeed, this must be the case when other acids are employed for the disengagement of carbonic acid gas, which do not, like the sulphuric, produce heat on being mixed with water. Carbonic acid may likewise be disengaged from its combinations by heat alone, which restores it to its gaseous state.

CAROLINE
It appears to me very extraordinary that the same gas, which is produced by the burning of wood and coals, should exist also in stones, marble and chalk, which are incombustible substances.

MRS. B.
I will not answer that objection, Caroline, because I think I can put you in a way of doing it yourself. Is carbonic acid combustible?

CAROLINE
Why, no— because it is a body that has been already burnt, it is carbone only, not the acid, which is combustible.

MRS. B.
Well and what inference do you draw from this?

CAROLINE
That carbonic acid cannot render the bodies in which it is contained combustible; but that simple carbone does, and that it is in this elementary state that it exists in wood, coals, and a great variety of other combustible bodies. — Indeed, Mrs. B., you are very ungenerous; you are not satisfied with convincing me that my objections are frivolous, but you oblige me to prove them to myself.

MRS. B.

You must confess, however, that I make ample amends for the detection of error, when I enable you to discover the truth.

CONVERSATION XVI

ON THE MURIATIC AND OXYGENATED MURIATIC ACIDS; AND ON MURIATS

Oxy-muriatic acid[1]

MRS. B.
....[*Oxy-muriatic acid gas*[2]] may be received over water, as it is but sparingly absorbed by it.— I have just collected some in this jar—

CAROLINE
It is not invisible, like the generality of gasses; for it is of a yellowish colour.

MRS. B.
The muriatic acid extinguishes flame, whilst, on the contrary, the oxy-muriatic makes the flame larger, and gives it a dark red colour. Can you account for this difference in the two acids?

EMILY
Yes, I think so; the muriatic acid cannot be decomposed, and therefore will not supply the flame with the oxygen necessary for its support; but when this acid is further oxygenated it will part with its additional quantity of oxygen, and in this way support combustion.

MRS. B.
That is exactly the case; indeed the oxygen, added to the muriatic acid, adheres so slightly to it, that it is separated by a mere exposure to the sun's rays. This acid is decomposed also by combustible bodies, many of which

---

[1] M2, 84-85.
[2] When Davy oxidised his *muriatic acid* gas (our *hydrogen chloride*, HCl), he obtained chlorine Cl_2, and rationally called it *oxy*-muriatic acid gas. As this contains no hydrogen, it can no longer be termed an acid in the sense of a substance which can split up to produce hydrogen ions.

it burns, and actually inflames, without any previous increase in temperature.

CAROLINE
That is extraordinary, indeed! I hope you mean to indulge us with some of these experiments?

MRS. B.
I have prepared several glass jars of oxy-muriatic acid gas, for that purpose. In the first we shall introduce some Dutch gold leaf[1]— Do you observe that it takes fire?

EMILY
Yes, indeed it does— how wonderful it is! it became instantly red hot, but was soon smothered into a thick vapour.

CAROLINE
Good heavens![2] what a disagreeable smell!

MRS. B.
We shall try the same experiment with phosphorus in another jar of this acid— you had better keep your handkerchief to your nose when I open it— now let us drop into it this little piece of phosphorus—

CAROLINE
It burns really; and almost as brightly as in oxygen gas! But, what is most extraordinary, these combustions take place, without the metal or phosphorus being previously lighted, or even in the least heated.

Muriat of ammonia[3]

MRS. B.
Muriat of ammonia is another combination of this acid, which we have already mentioned as the principal source from which ammonia is derived.

I can at once show you the formation of this salt by the immediate combination of muriatic acid with ammonia.— These two glass jars

---

[1] A thin bronze foil containing 11 parts of copper to 2 of zinc, which was used as a cheap substitute for gold.
[2] See p.51, note 2.
[3] M2, 91-92. Muriat of ammonia is our *ammonium chloride*, NH$_4$Cl.

contain, the one muriatic acid gas, the other ammoniacal gas, both of which are perfectly invisible— now, if I mix them together, you see they form immediately an opake white cloud, like smoke.— If a thermometer was placed in the jar in which these two gasses are mixed, you would perceive that some heat is at the same time produced.

EMILY
The effects of chemical combination are, indeed, wonderful— how extraordinary it is that two invisible bodies should become visible by their union!

MRS. B.
This strikes you with wonder, because it is a phenomenon which nature seldom exhibits to our view; but the most common of her operations are as wonderful, and it is their frequency only that prevents our regarding them with equal admiration. What would be more surprising, for instance, than combustion, were it not rendered so familiar by custom?

EMILY
That is true.— But pray, Mrs. B., is this white cloud the salt that produces ammonia? How different it is from the solid muriat of ammonia which you once showed us!

MRS. B.
It is the same substance which first appears in the state of vapour, but will soon be condensed, by cooling against the sides of the jar, in the form of very minute crystals.

Oxy-muriat of potash[1,2]

MRS. B.
If this salt be mixed, and merely rubbed together, with sulphur, phosphorus, charcoal or indeed any other combustible, it explodes strongly.....

I mean to show you this experiment, but I would advise you not to repeat it alone; for if care be not taken to mix only very small quantities at

---

[1] M2, 94-95.
[2] Confusingly, *oxy-muriat of potash* is not a salt of oxy-muriatic acid (elemental chlorine), but of chloric acid. Familiar as potassium chlorate $KClO_3$, it was formerly used as a weed-killer.

a time, the detonation will be extremely violent, and may be attended with dangerous effects[1]. You see I mix an exceedingly small quantity of the salt with a little powdered charcoal, in this Wedgwood mortar, and rub them together with a pestle—

CAROLINE
Heavens! How can such a loud explosion be produced by so small a quantity of matter?

MRS. B.
You must consider that an extremely small quantity of solid substance may produce a very great volume of gasses; and it is the sudden evolution of these which occasions the sound.

EMILY
Would not oxy-muriat of potash make stronger gunpowder than nitrat of potash?

MRS. B.
Yes; but the preparation, as well as the use of this salt, is attended with so much danger, that it is never employed for that purpose.

CAROLINE
There is no cause to regret it, I think; for the common gunpowder is quite sufficiently destructive.

MRS. B.
I can show you a very curious experiment with this salt; but it again must be on condition that you never attempt to repeat it by yourselves[2]. I throw a small piece of phosphorus into this glass of water; then a little oxy-muriat of potash; and, lastly, I pour in (by means of this funnel, so as to bring it in contact with the two other ingredients at the bottom of the glass) a small quantity of sulphuric acid –

CAROLINE
This is, indeed, a beautiful experiment! The phosphorus takes fire and burns from the bottom of the water.

---

[1] Mrs. B.'s condition is to be strictly observed. **ON NO ACCOUNT** try this experiment…
[2] …nor this one.

EMILY

How wonderful to see flame bursting out under water, and rising through it! Pray, how is it accounted for?

MRS. B.

Cannot you find it out, Caroline?

EMILY

Stop— I think I can explain it. Is it not because the sulphuric acid decomposes the salt by combining with the potash, so as to liberate the oxy-muriatic acid gas by which the phosphorus is set on fire?

MRS. B.

Very well, Emily; and with a little more reflection you would have discovered another concurring circumstance, which is, that an increase in temperature is produced by the mixture of the sulphuric acid and water, which assists in promoting the combustion of the phosphorus.

CONVERSATION XVII

ON THE NATURE AND COMPOSITION OF VEGETABLES

Organized bodies[1]

MRS. B.
It is now time to turn our attention to a more complicated class of compounds, that of ORGANIZED BODIES, which will furnish us with a new source of instruction and amusement.

EMILY
By organized body, I suppose, you means the vegetable and animal creation? I have, however, but a very vague idea of the word *organization*, and I have often wished to know more precisely what it means.

MRS. B.
Organized bodies are such as are endowed by nature with various parts, particularly constructed to perform certain functions connected with life. Thus you may observe, that mineral compounds are formed by the simple effect of mechanical or chemical attraction, and may appear to some to be, in a great measure, the productions of chance; whilst organized bodies bear the most striking and impressive marks of design, and are eminently distinguished by that unknown principle, called *life*, from which the various organs derive the power of exercising their respective functions.

CAROLINE
But in what manner does life enable these organs to perform their several functions?

---

[1] M2, 97-99.

MRS. B.

That is a mystery which, I fear, is enveloped in too profound darkness for us to hope that we shall ever be able to unfold it. We must content ourselves with examining the effects of this principle; as for the cause, we have been able only to give it a name, without attaching any other meaning to it than the vague and unsatisfactory idea of an unknown agent.

CAROLINE

And yet I think I can form a very clear idea of life.

MRS. B.

Pray let me hear how you would define it?

CAROLINE

It is perhaps more easy to conceive than to express— let me consider— Is not life the power which enables both the animal and the vegetable creation to perform the various functions which nature has assigned to them?

MRS. B.

I have nothing against your definition; but you will allow me to observe, that you have only mentioned the effects which the unknown cause produces, without giving us any notion of the cause itself.

EMILY

Yes, Caroline, you have told us what life *does*, but you have not told us what it *is*.

MRS. B.

We may study its operations, but we should puzzle ourselves to no purpose by attempting to form an idea of its real nature.

Sugar[1]

MRS. B.
Sugar is not found in its simple state in plants, but is always mixed with gum, sap, or other ingredients; it is to be found in every vegetable, but abounds most in roots, fruits, and particularly in the sugar cane.

EMILY
If all vegetables contain sugar, why is it extracted from the sugar-cane?

MRS. B.
Because it is most abundant in that plant, and most easily extracted from it.

During the late troubles in the West Indies[2], when Europe was but imperfectly supplied with sugar, several attempts were made to extract it from other vegetables, and very good sugar was obtained from parsnips and from carrots; but the process was too expensive to carry this enterprise to any extent[3].

CAROLINE
I should think that sugar might be more easily obtained from sweet fruits, such as figs, dates &c.

MRS. B.
Probably; but it would be still more expensive, from the high price of those fruits.

EMILY
Pray, in what manner is sugar obtained from sugar-cane?

---

[1] M2, 106-107.
[2] In the 1790's, there were serious insurrections among the black slaves who worked the sugar plantations; and in Santo Domingo (now Haiti) they had won their freedom by the end of the decade. But the uprisings there continued and in 1801-2 attempts to subdue them, together with an outbreak of yellow fever, cost the French Army 50,000 lives.
[3] I am most grateful to Katherine Cotter, of the British Beet Research Organisation, for telling me that sugar was first extracted from the root of sugar beet in Germany, as early as 1748, the first sugar-beet factory being opened in Breslau in 1799. After the onset of the Napoleonic wars (1803-15) sugar from beet replaced cane-sugar in France where, by 1813, over 300 sugar-beet factories were operation. Jane Marcet was probably not the only Londoner to be unaware of beet-sugar in 1805; and there was no sugar-beet factory in England until 1912.

MRS. B.

The juice of this plant is first expressed by passing it between two cylinders of iron. It is then boiled with lime-water, which makes a thick scum rise to the surface. The clarified liquor is let off below and evaporated to a very small quantity, after which it is suffered to crystallize by standing in a vessel, the bottom of which is perforated with holes, that are imperfectly stopped, in order that the syrup may drain off. The sugar obtained by this process is a coarse brown powder, commonly called raw or moist sugar; it undergoes another operation to be refined and converted into loaf sugar. For this purpose it is dissolved in water, and afterwards purified by an animal fluid called albumin. White of egg chiefly consists of this fluid, which is also one of the constituent parts of blood; and consequently eggs, or bullocks' blood, are commonly used for this purpose.

The albuminous fluid being diffused through the syrup, combines with all the solid impurities contained in it, and rises with them to the surface, where it forms a thick scum; the clear liquor is the again evaporated to a proper consistence, and poured into moulds, in which, by a confused crystallization, it forms loaf sugar. But an additional process is required to whiten it; to this effect the mould is inverted, and its open base covered with clay, through which water is made to pass; the water slowly trickling through the sugar, combines with and carries off the colouring matter.

CAROLINE

I am very glad to hear that the blood that is used to purify sugar does not remain in it; it would be a disgusting idea.

EMILY

And pray how is sugar-candy and barley-sugar prepared?

MRS. B.

Candied sugar is nothing more than the regular crystals, obtained by slow evaporation from a solution of sugar. Barley sugar is sugar melted by heat, and afterwards cooled in moulds of spiral form.

Sugar may be decomposed by red heat, and, like all other vegetable substances, resolved into carbonic acid and hydrogen.

Camphor[1]

MRS. B.

Among the particular properties of camphor, there is one too singular to be passed over in silence. If you take a small piece of camphor, and place it on the surface of a bason of pure water, it will immediately begin to move round and round with great rapidity; but if you pour into the bason a single drop of any odoriferous fluid, it will instantly put a stop to this motion. You can at any time try this very simple experiment[2]; but you must not expect that I shall be able to account for this phenomenon, as nothing satisfactory has yet been advanced for its explanation.

Tannin[3]

[*Mrs. B. has just explained that the tannin which is used to convert skins to leather occurs not only in oak bark, but also in wine*].

CAROLINE

One might suppose that men who drink large quantities of red wine, stand a chance of having the coats of their stomachs converted into leather, since tannin has so strong an affinity for skin.

MRS. B.

It is not impossible but that the coats of their stomachs may be, in some measure, tanned, or hardened, by constant use of this liquor; but you must remember that where a number of other chemical agents are concerned, and, above all, where life exists, no certain chemical inference can be drawn.

I must not dismiss this subject, without mentioning a very recent discovery of Mr. Hatchett[4] which relates to it. This gentleman found that a substance very similar to tannin, possessing all its leading properties, and actually capable of tanning leather, may be produced by exposing carbone, of any substance containing carbonaceous matter, whether vegetable, animal, or mineral, to the addition of nitric acid.

---

[1] M2, 119.
[2] The experiment is particularly entertaining if the camphor is placed in the stern of a tiny tinfoil boat, a familiar toy in the 1930's.
[3] M2, 127-129.
[4] C. Hatchett (1765-1847), of London, was a chemical analyst and manufacturing chemist.

CAROLINE
And is not this discovery likely to be of great use to manufactures?

MRS. B.
That is very doubtful, because tannin, thus artificially prepared, must probably always be more expensive than that obtained from bark. But the fact is extremely curious, as it affords one of those very rare instances of chemistry being able to imitate the proximate principles of organized bodies.

CONVERSATION XVIII

ON THE DECOMPOSITION OF VEGETABLES

The combustion of alcohol[1]

EMILY
I suppose that alcohol must be highly combustible, since it contains so large a proportion of hydrogen?

MRS. B.
Extremely so; and it will burn at a very moderate temperature.

CAROLINE
I have often seen both brandy and spirit of wine burnt; they produce a great deal of flame, but not a proportional quantity of heat, and no smoke whatever.

MRS. B.
The last circumstance arises from their combustion being complete; and the disproportion between the flame and the heat shows you that these are by no means synonymous.

The great quantity of flame proceeds from the combustion of the hydrogen to which, you know, that manner of burning is peculiar.— Have you not remarked also that brandy and alcohol will burn without a wick?— They take fire at so low a temperature, that this assistance is not required to concentrate the heat and volatilize the fluid.

CAROLINE
I have sometimes seen brandy burnt merely by heating it in a spoon.

MRS. B.
The rapidity of the combustion of alcohol may, however, be prodigiously increased by first volatilizing it. An ingenious instrument has been constructed on this principle to answer the purpose of a blow-pipe,

---

[1] M2, 157-161.

which may be used for melting glass or other chemical processes. It consists of a small metallic vessel (PLATE X. Fig.24), of spherical shape, which contains the alcohol, and is heated by the lamp beneath it; as soon as the alcohol is volatilized, it passes through the spout of the vessel, and issues just above the wick of the lamp, which immediately sets fire to the stream of vapour, as I shall show you—

EMILY

With what amazing violence it burns! The flame of the alcohol, in the state of vapour, is, I fancy, much hotter than when the spirit is merely burnt in a spoon?

MRS. B.

Yes; because in this way the combustion goes on much quicker, and, of course, the heat is proportionally increased. — Observe its effect on this small glass tube, the middle of which I present to the extremity of the flame, where the heat is greatest.

CAROLINE

The glass, in that spot, is become red hot, and bends from its own weight.

MRS. B.

I have now drawn it asunder, and am going to blow a bulb at one of the heated ends; but I must previously close it up, and flatten it with this little metallic instrument, otherwise the breath would pass through the tube without dilating any part of it.— Now, Caroline, will you blow strongly into the tube whilst the closed end is still red hot.

EMILY

You blowed too hard; for the ball suddenly dilated to a great size, and then burst into pieces.

MRS. B.

You will be more expert another time; but I must caution you, should you ever use this blow-pipe, to be very careful that the combustion of the alcohol done not go on with too great violence, for I have seen the flame sometimes dart out with such force as to reach the opposite wall of the room, and set the paint on fire. There is, however, no danger of the vessel bursting, as it is provided with a safety tube, which affords an additional vent for the vapour of alcohol when required.

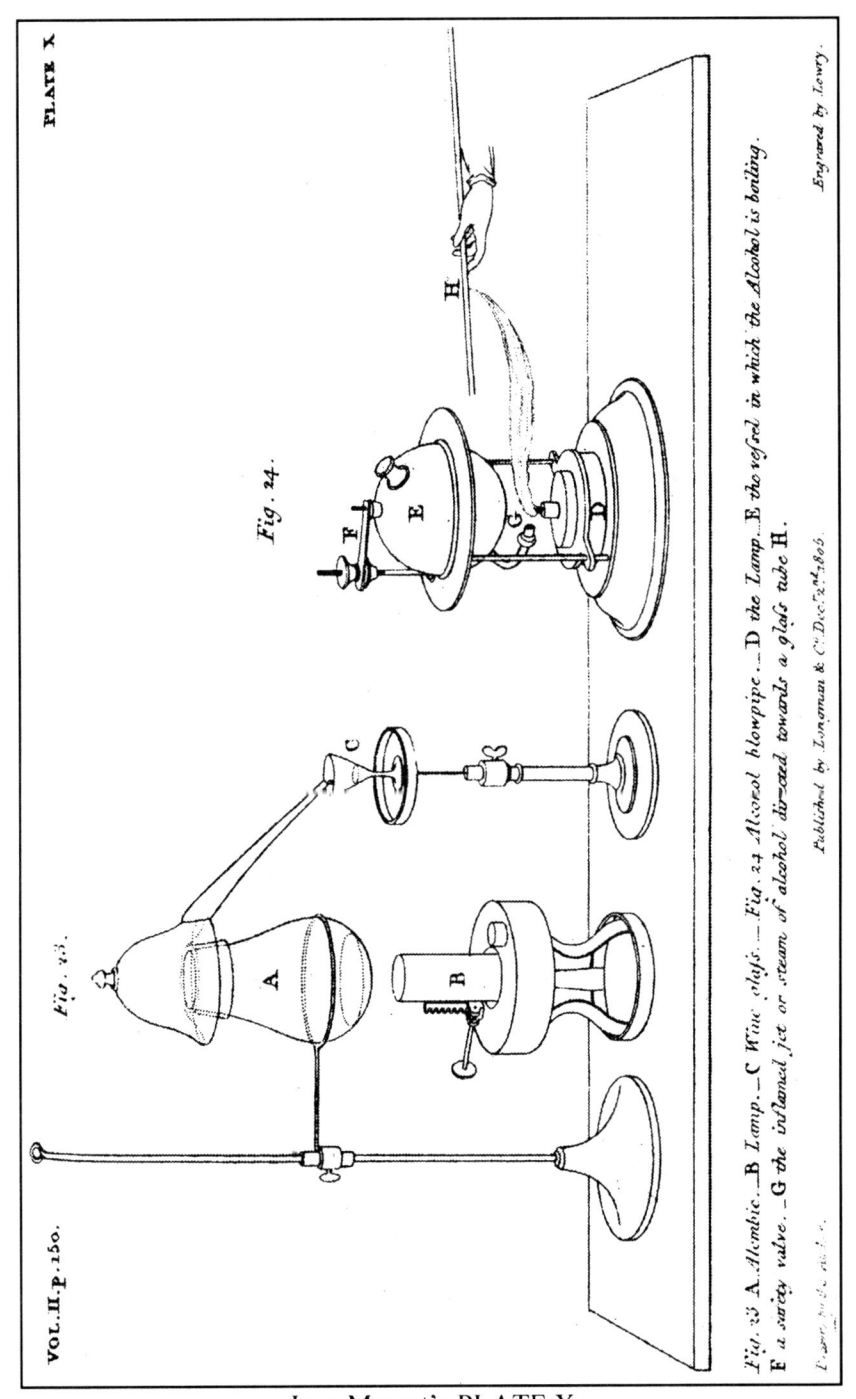

Jane Marcet's PLATE X

The products of combustion of alcohol consist in a great proportion of water, and a small quantity of carbonic acid. There is no smoke or fixed remains whatever. — How do you account for that, Emily?

EMILY

I suppose that the oxygen which the burning alcohol absorbs in burning, converts its hydrogen into water and its carbone into carbonic acid gas; and thus is completely consumed.

MRS. B.

Very well.

CONVERSATION XIX

HISTORY OF VEGETATION

Vegetation around London[1]

CAROLINE
Perhaps the smoky atmosphere of London *[being rich in carbone, a component of plants]* is the reason why the vegetation is so forward and so rich in its vicinity?

MRS. B.
I rather believe that this circumstance proceeds from the very ample supply of manure, assisted perhaps by the warmth and shelter which the town affords. Far from attributing any good to the smoky atmosphere of London, I confess I like to anticipate the time when we shall have made such progress in the art of managing combustion that every particle of carbone will be consumed, and the smoke destroyed, at the moment of its production. We may then expect to have the satisfaction of seeing the atmosphere of London as clear as that of the country.

Agriculture and manufacture[2]

EMILY
I have often thought that the culture of land was not considered as a concern of sufficient importance. Manufactures always take the lead; and health and innocence are often sacrificed to the prospect of a more profitable employment. It has often grieved me to see the poor manufacturers crowded together in close rooms, and confined for the whole day to the most uniform and sedentary employment, instead of

---

[1] M2, 183-184.
[2] M2, 184-186.

being engaged in that innocent and salutary kind of labour, which nature seems to have assigned to man for the immediate acquirement of comfort, and for the preservation of his existence. I am sure that you agree with me in thinking so, Mrs. B.?

MRS. B.

I am entirely of your opinion, my dear, in regard to the importance of agriculture; but I am far from wishing to depreciate manufactures; for as the labour of one man is sufficient to produce food for several, those whose industry is not required in tillage must do something in return for the food that is provided for them. They exchange, consequently, the accommodations for the necessaries of life. Thus the carpenter and the weaver lodge and clothe the peasant, who supplies them with their daily bread. The greater the stock of provisions, therefore, which the husbandman produces, the greater is the quantity of accommodation which the artificer prepares. Such are the happy effects which naturally result from civilized society. It would be wiser, therefore, to endeavour to improve the situation of those who are engaged in manufactures, than to indulge in vain declamations on the hardships to which they are often exposed.

Water, photosynthesis and respiration[1]

MRS. B.

I have said that water forms the chief nourishment of plants; it is the basis not only of the sap, but of all the vegetable juices. Water is the vehicle which carries into the plant the various salts and other ingredients required for the formation and support of the vegetable system. Nor is this all; the great part of the water itself is decomposed by the organs of the plant; the hydrogen becomes a constituent part of oil, of extract, of colouring matter &c, whilst a portion of the oxygen enters into the formation of mucilage, of fecula[2], of sugar, and of vegetable acids. But the greater part of the oxygen, proceeding from the decomposition of the water, is converted into a gaseous state by caloric disengaged from the hydrogen during its condensation in the formation of the vegetable materials. In this state the oxygen is transpired by the leaves of plants when exposed to the sun's rays. Thus you find that the decomposition of water, by the organs of the plants, is not only a means of supplying it with

---

[1] M2, 192-195.
[2] Formerly *faecula*, i.e. sediment from vegetable matter, e.g. lees of wine.

its chief ingredient, hydrogen, but at the same time of replenishing the atmosphere with oxygen, a principle which requires continual renovation, to make up for the great consumption of it occasioned by the numerous oxygenations, combustions, and respirations, that are constantly taking place on the surface of the globe.

EMILY

What a striking instance of the harmony of nature!

MRS. B.

And how admirable the design of Providence, who makes every different part of the creation thus contribute to the support and renovation of each other!

But the intercourse of the vegetable and animal kingdoms extends still further. Animals, in breathing, not only consume the oxygen of the air, but load it with carbonic acid, which, if accumulated in the atmosphere, would, in a short time, render it totally unfit for respiration. Here the vegetable kingdom again interferes; it attracts and decomposes the carbonic acid, retains the carbone for its own purposes, and returns the oxygen for ours. This process, however, is only carried on during the day, and a contrary one seems to take place at night; for the leaves then absorb oxygen and emit carbonic acid. The absorption of carbonic acid during the day is, however, far from balanced by the quantity emitted during the night.

CAROLINE

How interesting this is! I do not know a more beautiful illustration of the wisdom which is displayed in our laws of nature.

MRS. B.

Faint and imperfect as are the ideas which our limited perceptions enable us to form of divine wisdom, still they cannot fail to inspire us with awe and admiration. What then would be our feelings were the complete system of nature at once displayed before us! So magnificent a scene would probably be too great for our limited and imperfect comprehension, and it is no doubt, amongst the wise dispensations of Providence, to veil the splendour of a glory with which we should be overpowered. But it is well suited to a rational being to explore, step by step, the works of the creation, to endeavour to connect them into harmonious systems; and, in a word, to trace, in the chain of beings, the kindred ties and benevolent design which unite its various links, and secure its preservation.

CONVERSATION XX

ON THE COMPOSITION OF ANIMALS

Elements in nature[1]

CAROLINE
Is it not surprising that so great a variety of substances, and so different in their nature, should yet all arise from so few materials, and from the same original elements?

MRS. B.
The difference in the nature of various bodies depends, as I have often observed to you, rather on their state of combination, than on the materials of which they are composed. Thus, in considering the nature of the creation in a general point of view, we observe that it is throughout composed of a very small number of elements. But when we divide it into the three kingdoms, we find that, in the mineral, the combinations seem to result from the union of elements casually brought together; whilst in the vegetable and animal kingdoms, the attractions are peculiarly and regularly produced by appropriate organs, whose action depends on the vital principle. And we may further observe, that by means of certain spontaneous changes and decompositions, the elements of one kind of matter become subservient to the reproduction of another; so that the three kingdoms are intimately connected, and constantly contributing to the preservation of each other.

EMILY
There is, however, one very considerable class of elements, which seems to be confined to the mineral kingdom: I mean metals.

MRS. B.
Not entirely; they are found, though in very minute quantities, both in the vegetable and animal kingdoms. A small portion of earths and sulphur also enters into the composition of organized bodies. Phosphorus,

---
[1] M2, 206-207.

however, is almost entirely confined to the animal kingdom; and nitrogen, with but few exceptions, is extremely scarce in vegetables.

The diversity of chemistry[1]

EMILY
And when jelly is made of isinglass[2] does it leave no sediment?

MRS. B.
No; nor does it so much require clarifying, as it consists almost entirely of pure gelatine, and any foreign matter that is mixed with it, is thrown off during the boiling in the form of scum.— These are processes which you may see performed in great perfection in the culinary laboratory, by that very able and most useful chemist the cook.

CAROLINE
To what an immense variety of circumstances chemistry is subservient!

EMILY
It appears, in that respect, to have an advantage over most other arts and sciences; for these, very often, have a tendency to confine the imagination to their own particular object, while the pursuit of chemistry is so extensive and diversified, that it inspires a general curiosity, and a desire of inquiring into the nature of every object.

Prussic acid[3]

EMILY
But how can prussic acid be artificially made?

---

[1] M2, 212.
[2] Impure gelatin, obtained from the air-sacs of fish and other sources.
[3] M2, 218-219.

MRS. B.

By passing ammoniacal gas over red hot charcoal; and hence we learn that the constituents of this acid are hydrogen, nitrogen and carbone. The first two are derived from the volatile alkali, and the last from the combustion of the charcoal.

CAROLINE

But this does not accord with the system of oxygen being the indispensable principle of acidity?

MRS. B.

It is true; and this circumstance, together with some others of the same kind, has led several chemists to suspect that oxygen may not be the sole generator of acids[1], and that acidity may possibly depend rather on the arrangement than on the presence of any particular principles.

CAROLINE

I do not like the idea. For if it were founded, all our theory of chemistry must be erroneous.

MRS. B.

The objection is yet so new and unconfirmed by common experience, that I confess I do not feel inclined to distrust the general doctrine of acidification which we have hitherto adopted.

---

[1] See pp. 107, 148.

CONVERSATION XXI

ON THE ANIMAL ECONOMY

Bodily exercise[1]

MRS. B.
Exercise is generally beneficial to all the animal functions. If man is destined to labour for his subsistence, the bread which he earns is scarcely more essential to his health and preservation than the exertions by which he obtains it. Those whom the gifts of fortune have placed above the necessity of bodily labour, are compelled to take exercise in some mode or other, and when they cannot convert it into an amusement, they must submit to it as a task, or their health will soon experience the effects of their indolence.

EMILY
That will never be my case; for exercise, unless it becomes fatigue, always gives me pleasure; and, so far from being a task, is to me a source of daily enjoyment. I often think what a blessing it is, that exercise, which is so conducive to health, should be so delightful; whilst fatigue, which is rather hurtful, instead of pleasure, occasions painful sensations. So that fatigue, no doubt, was intended to modify our bodily exertions, as satiety puts a limit to our appetites?

MRS. B.
Certainly.— But let us not deviate too far from our subject. —

---

[1] M2, 228.

CONVERSATION XXII

ON ANIMALIZATION, RESPIRATION, AND NUTRITION

Respiration[1]

MRS. B.

[Respiration] is one of the grand mysteries which modern chemistry has disclosed. When the venous blood enters the left ventricle of the heart, it contracts by its muscular power, and throws the blood through a large vessel into the lungs, which are contiguous, and through which it circulates by millions of small ramifications. Here it comes into contact with the air we breathe. The action of the air or the blood in the lungs is, indeed, concealed from our immediate observation; but we are able to form a tolerably accurate judgment of it from the changes which it effects not only in the blood, but also on the air expired.

This air is found to contain all the nitrogen inspired, but to have lost part of its oxygen, and to have acquired a portion of watery vapour. Hence it is inferred, that when the air comes in contact with the venous blood in the lungs, the oxygen attracts from it the superabundant quantity of hydrogen and carbone with which it has impregnated itself during the circulation; and that one part of that oxygen combines with the hydrogen, in the form of watery vapour, whilst another part combines with the carbone, which it converts into carbonic acid. The whole of these products being then expired, the blood is restored to its former purity, that is to the state of arterial blood, and is thus again enabled to perform its various functions.

CAROLINE

This is truly wonderful! Of all the things we have yet learned, I do not recollect any thing that has appeared to me so curious and interesting. I almost believe I should like to study anatomy now, though I have hitherto

---

[1] M2, 252-253 and 257-259.

had so disgusting an idea of it. Pray, to whom are we indebted for these beautiful discoveries?

MRS. B.

Crawford[1], in this country, and Lavoisier, in France, are the principal inventors of the theory of respiration. But the still more important and more admirable discovery of the circulation of the blood was made long before by our immortal countryman, Harvey[2].

EMILY

Indeed I have never heard of any thing that delighted me so much as this theory of respiration.......

MRS. B.

....But before we proceed further, I must inform you that the chemical theory of respiration, with which you have just been made acquainted, simple and beautiful as it is, has appeared to many philosophers insufficient to explain all the phenomena of respiration. Amongst the various modifications proposed, with a view to improve this theory, that suggested by La Grange[3], Hassenfratz[4], and some other eminent chemists, appears to be the most important. These gentlemen suppose that the oxygen, which disappears in respiration, is absorbed by the blood, and carried with it into the circulation, during which it gradually combines with the hydrogen and carbone that are successively added to the circulation, forming water and carbonic acid which are expelled from the lungs at each expiration. Thus the process, instead of being completed in the lungs, as the former theory supposes, only begins in that organ, and continues throughout the whole circulation.

According to this theory, the florid colour of arterial blood depends on the addition of oxygen, so that this colour gradually vanishes as the blood passes from the arterial to the venous state, that is to say, as the oxygen enters into combination with the hydrogen and carbone during circulation.

---

[1] A. Crawford (1748-95), Irish physician and chemist.
[2] W. Harvey (1578-1657).
[3] La Grange, referred to by Hassesnfratz as de la Grange, is almost certainly J. L. Lagrange (1736-1813), a mathematician and astronomer who also did chemical experiments. A friend of Lavoisier's, he is quoted as saying on the day after Lavoisier was guillotined *"only a moment was needed to cause that head to fall, and a hundred years will perhaps not suffice to raise another like it."*
[4] J-H. Hassenfratz (1755-1827), Deputy Inspector of Mines, worked with Lavoisier. See *Annales de Chimie,* **9** (1791) 261-74.

CAROLINE

There does not appear to me to be any very essential difference in these two theories, since in both the oxygen purifies the blood by combining with and carrying off the matter which had accumulated in it during circulation.

MRS. B.

Yes; but, in medical, or rather physiological science, it must be a question of great importance whether the oxygen actually enters the circulation, or whether it proceeds no further than the lungs.

CONVERSATION XXIII

ON ANIMAL HEAT; AND ON VARIOUS ANIMAL PRODUCTS

Body temperature[1]

MRS. B.
I should think that the principal cause of the heat experienced in fevers, is, that there is no vent for the caloric which is generated in the body. One of the most considerable secretions is perspiration; this is constantly carrying off caloric in a latent state; but during the hot stage of a fever, the pores are so contracted that all perspiration ceases, and the accumulation of caloric in the body occasions those burning sensations which are so painful.

EMILY
This is, no doubt, the reason why the perspiration that often succeeds the hot stage of a fever affords so much relief. If I had known this theory of animal heat when I had a fever last summer, I think I should have found some amusement in watching the chemical processes that were going on within me.

CAROLINE
But exercise likewise produces animal heat, and that must be in quite a different manner.

MRS. B.
Not so much as you think; for the more exercise you take, the more the body is stimulated, and requires recruiting. For this purpose the circulation of the blood is quickened, the breath proportionately accelerated, and consequently a greater quantity of caloric evolved.

---

[1] M2, 266-269.

CAROLINE

True; after running very fast, I gasp for breath, my respiration is quick, and hard, and it is just then that I begin to feel hot.

EMILY

It would seem, then, that violent exercise should produce fever.

MRS. B.

Not if the person is in a good state of health; for the additional caloric is then carried off by the additional caloric which succeeds.

EMILY

What admirable resources nature has provided for us! By the production of animal heat she has enabled us to keep up the temperature of our bodies above that of inanimate objects; and whenever this source becomes too abundant, the excess is carried off by perspiration.

MRS. B.

It is by the same law of nature that we are enabled, in all climates, and in all seasons, to preserve our bodies of an equal temperature, or at least very nearly so.

CAROLINE

You cannot mean that our bodies are of the same temperature in summer and in winter, in England, and in the West Indies?

MRS. B.

Yes, I do; at least if the speak of the temperature of the blood, and the internal parts of the body; for those parts which are immediately in contact with the atmosphere, such as the hands, and face, will occasionally get warmer, or colder, than the internal or more sheltered parts. If you put the bulb of a thermometer in your mouth, which is the best way of ascertaining the real temperature of your body, you will scarcely perceive any difference in its indication, whatever may be the difference of temperature of the atmosphere.

CAROLINE

And when I feel overcome by heat, am I really not hotter than when I am shivering with cold?

MRS. B.

When a person in health feels very hot, whether from internal heat, from violent exercise, or from the temperature of the atmosphere, his body is certainly a little warmer than when he feels very cold; but this difference is much smaller than our sensations would make us believe; and the natural standard is soon restored by rest and by perspiration. I am sure that you will be surprises to hear that the internal temperature of the body[1] scarcely ever descends below 95° or 96°, and hardly ever attains 104° or 105°, even in the most violent fevers.

Curds and whey[2]

[Mrs. B. has explained that alkalies, and rennet, as well as acids, wine and spirituous liquors are able to separate milk curds from whey]

EMILY

This is a very useful piece of information; for I find white-wine whey, which I sometimes take when I have a cold, extremely heating; now, if the whey were separated by means of an alkali instead of wine, it could not produce that effect.

MRS. B.

Perhaps not. But I would strenuously advise you not to place too much reliance on your slight chemical knowledge in medical matters. I do not know why whey is not separated from curd by rennet, or by an alkali, for the purpose which you mention; but I strongly suspect that there must be some good reason why the preparation by means of wine is generally preferred.

Finis[3]

MRS. B.

A manufacture has in consequence *[of the fact that decaying animal matter may be put to good use]* been established near Bristol, in which, by

---

[1] Corresponding to a range of 35 to 40.5°C.
[2] M2, 282.
[3] M2, 288-289.

exposing the carcasses of horses and other animals for a length of time under water, the muscular parts are converted into this spermaceti-like substance. The bones afterwards undergo a different process to produce hartshorn, or, more properly, ammonia, and phosphorus; and the skin is prepared for leather.

Thus art contrives to enlarge the sphere of useful purposes, to which the elements were intended by nature; and the productions of the several kingdoms are frequently arrested in their course, and variously modified, by human skill, which compels them to contribute, under new forms, to the necessities or luxuries of man.

But all that we enjoy, whether produced by the spontaneous operations of nature, or the ingenious efforts of art, proceed alike from the goodness of Providence.— .To GOD alone man owes the admirable faculties which enable him to improve and modify the productions of nature, no less than those productions themselves. In contemplating the works of the creation, or studying the inventions of art, let us, therefore, never forget the Divine Source from which they proceed; and thus every acquisition of knowledge will prove a lesson in piety and virtue.

APPENDICES

I A contemporary review

An anonymous reviewer[1] deals first with the novelty of a chemistry book written by a lady and the suitability of scientific knowledge as a feminine accomplishment:

> *The present publication is the work of a lady, who, as far as we know, is the first female who has ever favoured the world with any thing like a system of chemistry. Now, although scientific knowledge is not exactly what the eye looks for and the heart desires in women, yet it cannot be said to be incompatible either with the more showy or fashionable of female accomplishments, or with the more important of female duties. Many people, however, are of a different opinion, and will tell you that the female mind is not endowed with a capacity for scientific investigation, and if it is, that it is inexpedient to cultivate that capacity. Perhaps our ancestors were of that opinion. In their public institutions for the advancement of science there was no room left for the admission of females. But moderns think and act differently in that respect.*

He draws heavily on the Preface to account for the author's reasons for writing the book, and continues:

> *The design was certainly laudable. Has it been executed with success?*

There follows a detailed account of the contents of the book with a few adverse comments. Posterity would certainly support the author against some of the reviewer's complaints, such as his objection that the atmosphere, separable as it is into oxygen and nitrogen, should be treated as a Compound Body. On the other hand, we might agree with him that *the term chemical heat is ... objectionable, because it seems to suggest a separate species of heat, while in fact it denotes only a particular modification.* But we would surely not uphold his contention that there is no well-founded distinction between the solution of salt in water, and that

---

[1] *The Literary Journal*, Vol. I, Second Series, p.180 (London, 1806).

of metal in acid. The bulk of the review is, however, descriptive rather than critical; and it ends as follows:

Upon the whole it is a book which we have no hesitation in recommending to all such as are entering upon the study of chemistry, and who wish to have the useful mixed with a little of the sweet. They will find Mrs. B. to be a very intelligent instructor, and Emily and Caroline to be very attentive pupils. The dialogue is in general lively and spirited, and if it is but seldom elevated, it is never flat.

II *Michael Faraday's appreciation*[1]

After Jane Marcet's death, on June 28, 1858 her children asked Auguste de la Rive to write a short biography of her. On 27 July, he wrote to Michael Faraday in simple French asking if it was true that his reading of *Conversations on Chemistry* had first inspired his taste for chemistry and physics, and had determined the direction of his work; and if this were so, whether he might mention it in the obituary. Faraday replied as follows:

Hampton Court

2 October 1858

My Dear Friend,

   Your subject interests me deeply every way; for Mrs. Marcet was a good friend to me, as she must have been to many of the human race. I enterd the shop of a bookseller and bookbinder at the age of 13, in the year 1804, remained there 8 years, and during part of the time bound books. Now it was in these books, in the hours after work, that I found the beginnings of my philosophy. There were two that especially helped me; the Encyclopaedia Britannica, from which I gained my first notions of Electricity; and Mrs. Marcets conversations on chemistry, which gave me my foundation in that science. I believe I had read about phlogiston etc in the Encyclopaedia, but her book came as the full light in my mind. Do not suppose that I was a very deep thinker or was marked as a precocious

---

[1] See p. iv.

person;– I was a very lively, imaginative person, and could believe in the Arabian nights as easily as in the Encyclopaedia. But facts were important & saved me. I could trust a fact,– but always cross-examined an assertion. So when I questioned Mrs. Marcets book by such little experiments as I could find the means to perform, & found it true to the facts as I could understand them, I felt I had got hold of an anchor in chemical knowledge & clung fast to it. Hence my deep veneration for Mrs. Marcet; first as one who had conferred great personal good & pleasure on me;– and then as one able to convey the truths and principles of those boundless fields of knowledge which concern natural things to the young, untaught, and enquiring mind.

You may imagine my delight when I came to know Mrs. Marcet personally;– how often I cast my thoughts backward delighting to connect the past and the present;– how often when sending a paper to her as a thank offering I thought of my first instructress;– and such like thoughts will remain with me…………

*Ever My dear friend
Yours Affectionately
M. Faraday*

INDEX

A
acetic acid, 109
   see also Vinegar
acids, 57-9, 63
   antidotes for, 113-4
   classification of, 106-9
   as compounds, 106-7
   with metals, 80-4
   nomenclature of, 57-9, 69, 106-107,109, 128
   oxygen in, 57, 107, 109 128, 148
   poisoning, 113-4
   see also individual acids
aeriform state, 17
affinity, 9, 11, 56, 78, 103-5
   elective, 103-5
agriculture, 1, 143-4
air,
   exhaled, 37, 77, 101-2, 150
   purity of, 63-4
   see also Atmosphere, Gases
alchemists, vii, 2
alcohol, 139-142
Algarotti, x, xvi
alkalies, 1, 89-96, 114, 125
   volatile, 89-90, 94-6
   see also Potash, soda
alkaline earths, *see* Earths
alloys, 86-7, 129
alum, 99
alumine, 99-100
America, publication in, v-vi
ammonia, 89, 94-6, 121-2, 129-130, 148, 156
   nitrat(e) of, 121
ammonium chloride,130
analysis, 67
animal kingdom, 108, 136, 146, 149
animal heat, 153-5
animals, composition of, 146-7
   decay of, 155-6
animals,
   experimental, xi, 34, 124
antimony, 78
aqua fortis, 118
Aristotle, vii-viii, xxiv, 4
art, 2-3
   see also Manufacture
ashes, 91-2
astronomy, ii, vii, ix, xii
atmosphere, composition of, 28-37
   purity of, 63, 143
atomic theory, viii, 69
attraction, of cohesion, 7, 28-30, 196
   of composition, 7
   see also Chemical attraction
aurora borealis, 119

B
Barbauld, A.L., xi, 34
barytes, 89
beet, sugar, 135
Berthollet, 96
beryllium oxide, 6
bleaching, 114-5
blow-pipe, 77, 139-140
"bluestockings", ix
body temperature, 153-5
boiling, 17
boracic acid, 107
Boyle, R., vii-viii, xvi, 4-5, 115
blood,
   circulation of, 150-3
   in respiration, 150-2
   in sugar refining, 136
brandy, 139
brass, 86
breathing, *see* Respiration
brick, 100
bronze, 86, 129
bubbles, *see* Soap-bubbles

C

caloric, 6, 14-17, 22-30, 39, 62, 84, 126
   and combustion, 28-31
   free, 15
   nature of, 26-7
   see also Combustion
calorimeter, 63
camphor, 137
candle, combustion of, 73
carbon(e), 67-73, 142-3, 148
   combustion of, 68-73, 124, 126
   forms of, 94
   gaseous oxyd of, 69
   in gunpowder, 123
   see also, Diamond
carbon dioxide, *see* Carbonic acid
carbonat(e), of lime, 100-2, 125-6
   of potash, 91-2
carbonated hydrogen gas, 72
carbonated hydrogen gas, 72
carbonic acid, 68-74, 108, 124-7, 136, 142, 144-5, 152
Cavendish, H., viii, xiii, 46, 117
cement, 98
ceramics, 98-100
chalk, 100-2, 125-6
change of state, 17, 53-6
charcoal,
   to decompose water, 42, 71-2
   and potassium chlorate, 131
   see also Blow-pipe, Carbone(e)
chemical attraction, 7-9, 85-6, 95, 103-5
   laws of, 103-5
   chemical elements, viii
   see also individual elements, Four Elements Hypothesis
chemical heat, *see* Heat, chemical
chemistry, benefits to society, xxi, 3
   development of, vii, xiii-xiv, 2, 4-5
   scope of, xix, xxv, 1-4, 147
chloric acid, 130
chlorine, 128-9
clay, 99

coal, combustion of, 36, 126
cobalt, 86
coldness, sensation of, 22-3
colour, 12-13
   of blood, 151
   of bubbles, 51
   of flames, 78
   of gems, 97
combustion, 40-53, 130, 143
   of alcohol, 139-142
   conditions for, 31
   of candle, 48, 72-4
   of carbon(e), 68-73, 126
   of coal, 31, 126
   of copper, 77
   of hydrogen, 43-8, 51-2
   of iron, 34-6, 77
      in nitrous oxide, 122
   of oil, 72-4
   oxygen in, 28-36
   of phosphorus, 62-3
   of sulphur, 56-7
   of tin, 77
   without oxygen, 129
   of wood, 28-31, 126
   see also Fire Triangle
compound bodies, 5, 90
compound salts, 82, 99-100
conduction, of heat, 23
constituent parts, 7, 10
conversations, *see* Dialogues
"*Conversations on Natural Philosophy*",
   xv, xvii-xviii, xxiv
cooking, 147
copper, 7-8, 76, 78, 81, 86
   in alloys, 86
   combustion of, 78
cotton, 67
Crawford, A., 151
crystallization, 84-5, 97
   of gems, 97
crystallization,
   water of, *see* Water

D

Dalton, J., viii, 69
Darwin, E., xxiv
Davy, H., iii, v, xxii, xxiv, 69, 90,
    93, 120, 128
decomposition, 10-11, 85
    see also, Water,
        decomposition of
density, 17
    of carbon dioxide, 125
detonation, *see* Explosion
dialogues, educational, xiii,
    xv-xxi, xxiv-xxv
diamond, 67-8, 93-3
dinitrogen monoxide,
    see Nitrous oxide
dissolution, 82
drugs, see Pharmacy

E

earths, 89, 97-102, 125-6,
    146
    see also Lime, magnesia
Edgeworth, M. xvii, xxv, 114
education, of boys, ii, xii
    of girls, ii viii, xi-xii
    scientific, vi, viii, xii
effervescence, 71,81
elastic fluids, 17
electric spark, 117-18
elementary bodies, 4-6
    see also Chemical elements,
    Four Element Hypothesis
Empedocles, viii, 4
eudiometer, 63
exercise, 149, 153, 155
exhalation, 36-7, 77
exhilarating gas,
    see Nitrous oxide
expansion, 16-8
explosion, 43-5, 48-52, 65,
    123-4, 131

F

Fahrenheit scale, 21-2
Faraday, M, .iv-v, 8, 158-9
Ferguson, J., xvii
fever, 153-5
fire, *see* Combustion
fire engine, 71-2
Fire Triangle, 31
fixed air, 124
fixed oils, 74
flint, 94, 98
fluoric acid, 107
Fontenelle, ix, xvi,
formic acid, 109
Four Element Hypothesis, viii, 4
Fourcroy, 92
French chemists, xiii-xiv, 105
    see also, Berthollet, Fourcroy,
    Lavoisier

G

Galileo, xvi
gases, early work on, xiii-xiv
gasses, *see also*, Aeriform state,
    gases
gelatine, 147
gems, 97
glass, 93-4, 98, 104
    blowing of, 140
Glauber's salt, 91-2, 116
gold, 2-3, 76
goldleaf, Dutch, 129
gunpowder, 122-4, 131

H

Haldimand, A., i-iii
Harris, J., xvi
Harrogate waters, 56-60
hartshorn, 98
Harvey, W., 151
Hatchett, C., 137
health and safety, xix, 66, 113, 121,
    131, 140, 143

heat, 12-18
   chemical, 15, 111-3, 116, 129-132
   conduction of, 23
   latent, 15, 24
   of reaction, *see* Heat, chemical
   specific, 15
   see also Caloric
Herschell, W., 12-4, 27
hotness, sensation of, 15, 22-3
hydrochloric acid, 107, 128
hydrocyanic acid, 109
hydrofluoric acid, 107
hydrogen, 39-52, 82, 121-2, 136, 142, 148
   in ammonia, 96
   in soap-bubbles, 48-52
   combined with oxygen, 40, 78-80
   combined with phosphorus, 55-6
   combined with sulphur, 59
   combustion of, 43-8, 51-2
   origin of name, 39, 92-2
   preparation of, 41-2

I
Ignes fatui, 65
inflammable air, 40
innovations, 25, 92-3
integrant parts, 9-11
iron,
   combustion of, 34-6, 78-80
   oxidation of 76, 78,
      see also Rust
   and sulphuric acid, 42, 83, 104
   tin-plated, 86

J
jelly, 147
Joyce, J., xxiv-xxv

L
labour, types of, 143-4
lactic acid, 109
Lagrange, J., 151
latent heat, 15, 24
laundering, 90
Lavoisier, A. xiii-xiv, xxi, 57, 86, 86, 90, 93, 151
lead, 76, 80, 86
leather, 137, 156
lemon juice, 80
life, principle of, 67
life sciences, *see* Living organisms, natural history, (and pp.131-156)
light, 6, 12-15, 27
lightning, 117-9
lime,
   and carbonic acid, 100
   carbonat(e) of, 100-2
   combined with phosphorus, 66,
lime water, 100-2
Linné, C., *see* Linnaeus, C.
Linnaeus, C., xii-xiii
liquid state, 17
living organisms, 67, 133-4, 137
Locke, J., vii, x, xii, xxiv
London, 143
lungs, *see* Air, exhaled

M
"Mad Madge", viii-ix
magnesia, 114
manganese, oxide of, 76
manufacture, 143-4
   of loaf sugar, 135-6
manure, 143
marble, 125
Marcet, A., i-iv, xxii, 47
Marcet, J., contribution of, xviii-xi
   early life, i-iii, iv-vi
   publications, iv-vi, 157-8
Martin, B., xvii
matches, 64, 115
medicines, *see* Pharmacy,

melting, 17
metals, 1, 75-86
   with acids, 80-4
   combustion of, 77-80
   in gems, 97
   in living systems, 146
   occurrence of, 75
   oxides of, 76
   see also individual metals
methanoic acid, 109
mineral acids, 109
mineral waters, 53, 59-60, 70-1
minerals, 67, 91, 99, 108, 146
mortar, 98
mouse, xi, 34
muriat of ammonia, 129-130
muriatic acid, 93, 95, 107, 128, 130

N

nature, 2-3, 67-8, 97, 130, 144-5, 154, 156
   imitation of, 67-8, 137-8
natural history, xii-xiii
natural philosophy,, xxiv, 1
   see also "Conversations in Natural Philosophy", Physics
Newton, I., vii-x
Nitrat(e), of ammonia, 121
   of potash, 118, 123, 131
nitre, 118, 123
nitric acid, 7-8, 81, 84, 104, 113, 117-18
nitrogen, 33-4, 37-8, 122, 147-8
   in air, 28, 33-4
   in ammonia, 96
   gaseous oxyd of, 120-2
   in living systems, 147
   insignificance of, 37-8
   see also Lightning
nitrous acid, 122
nitrous oxide, 120-2
nomenclature, chemical, xiii, xix, 57-9, 69, 86, 91-2, 95, 99, 106, 110-11, 118

Linnaean, xiii

O

oil, combustion of, 73
   in soap, 90-1, 93, 96
oil of vitriol, 110-11
   see also Sulphuric acid
oils, fixed, 74
optical instruments, vii, ix-xi, 13
organised bodies, 133-4
   see also Living organisms
oxidation, of metals, *see* Oxides
   acidic, 58
   of carbon(e), 69
      see also Carbonic acid
   of metals, 76, 78, 80
   of phosphorus, 62-4
   of sulphur, 56-57
   see also Water
oxyds, *see* Oxidation
oxygen, xiv
   in acids, 57-9
   in air, 28-36
   in acids, 57, 93
   combined with hydrogen, 40, 124
   in nature, 144-5
   in nitric acid, 117
   from nitrous oxide, 122
   origin of name, 57, 93
   in photosynthesis, 144-5
   preparation of, xiv, 76
   in respiration, 130, 150-2
   see also Combustion *(all entries)*, Lightning
"Oxygen", 33
oxymuriat of potash, 130-132
oxymuriatic acid, 128-129

P

perspiration, 153
pewter, 86
pharmacy, 2
Philosophers' Stone, (vii), 2

phosphine, 65-66
phosphorated hydrogen, 65-66
phosphoret of lime, 66
phosphoric acid, 108
phosphorus, 60-6, 129, 131-132, 156
   combined with hydrogen, 65-6
   combined with sulphur, 64
   combustion of, 62-3
   discovery of, 60-1
   in living systems, 61, 146
   in matches, 64, 115
photosynthesis, 144-5
phlogiston, 93
physics, 15
   see also Natural philosophy
plants, *see* Vegetable kingdom
platina, 76
Plato, xv
Pluche, A.N., xvii
poisoning, by acid, 113-14
pollution, by smoke, 143
popular science, ix-xii
porcelain, 100
potassium chlorate, 130-2
potash, 89-91
   carbonat(e) of, 91-2
   sulphat(e) of, 115-16
Priestley, J., xi-xiii, 34, 93, 120
principles, *see* Elements, chemical
Providence, 145, 156
Prussian blue, 109
prussic acid, 109, 147-8
pyrometer, 18

R
Réaumur scale, 20-1
respiration, 36-7, 77, 144-5, 150-152
Rousseau, J.-J., xi, xxiii
Royal Institution (London), iii, xiv, xxii
Royal Society (London), viii-ix

Rumford, Count, 25
rust, 76, 80, 83, 86-7

S
sal ammoniac, 95
saltpetre, 123
salts, 83
   compound, 83
sand, 94, 98-100
Scheele, 93
science,
   popular, ix-xii
   women in, ix-xi
Seltzer waters, 71-72, 101-2
silex, 98
silicious earth, 93
silver, 76
   see also Chemical elements
"singing flames", 47
smoke, pollution by, 143
soap, 90-1, 93, 96. 114
soap-bubbles, 48-52
society, 2-3
   and innovations, 25, 92-3
   working conditions in, 143-144
Socrates, xv
soda, 89-90
   sulphat(e) of, 116
soils, 99-100
solid state, 17
solution, types of, 85-6
Somerville, M., xxiv
Sophron, xv
specific heat, 15
spectrum, 12-13
states of matter, 17
steam engine, 3
steam heating, 24
strontites, 89
sublimation, 53-6
sugar, 135-6
sugar beet, 135
sulphat(e)s, *see* Potash, soda
sulphur, 53-64

combination with hydrogen, 59-60
combination with phosphorus, 64
combustion of, 56-7
flowers of, 53
in gunpowder, 123
in living systems, 53, 146
in matches, 64, 115
occurrence of, 53
sulphurated hydrogen, 59-60
sulphuric acid, 42, 58-9, 82, 104, 108, 110-14, 118
with carbonates, 125
concentrated, 110-13, 131-2
as poison, 113-14
sulphurous acid, 58-9, 114-15
"sympathetic ink", 87-8
synthesis, 67

T
tannin, 137
telescope, vii, xii, 13
temperature, of body, 153-5
and heat, 22
scales of, xix, 18-21
thermometers, 18-21, 154
tin, 77-78, 86
"Tom Telescope", x, xii, xvii
Thompson, B., *see* Rumford, Count
turpentine, 104

U
Unitarians, xii

V
vacuum, heat in, 26
vegetable kingdom, 67-8, 91, 108, 133-5, 143-6
vinegar, 80
vitriol, oil of, 110-14

W
water, 39-48, 144
decomposition of, 41, 71-2, 78-80
and hot coals, 71-2
as oxide, 40-1
as solvent, 85
of crystallization, 85, 121
and sulphuric acid, 121-2
West Indies, 135, 154
whey, 155
Will-of-the-Wisp, 65
wine, 137, 155
women, in science, x-xi, xiv-xviii, xxii, 2, 157,
see also Education
wood, combustion of, 28-31

Z
zinc, 78, 86

ABOUT THE EDITOR

Hazel Rossotti (née Marsh) was Fellow and Tutor in Chemistry at St. Anne's College, University of Oxford, UK for nearly forty years (and a part-time teacher of science to 12-year old boys for three). She studied for her first degree and her doctorate at Oxford University, where she married Francis, then a fellow research student. There followed research and teaching at Stockholm and Edinburgh, a break for family and book-writing, and return to Oxford. Although her main research interest is chemical equilibria in solution, she also enjoys writing about wider aspects of chemistry for a variety of readers, including young people and adult non-scientists.

Now a Senior Research Fellow of St. Anne's, she is exploring previous scientific books for pre-adult readers. She and her husband still live in Oxford, where she appreciates its libraries, townscape, countryside and good company; and its opportunities for continued learning: in stained-glass work, in modern and ancient Greek, in IT and, until recently, in windsurfing. But her main leisure interest is black-and-white photography.

Her previous books include:

for chemists:
The Determination of Stability Constants, with F.J.C. Rossotti (McGraw-Hill 1961, and Russian trans.).
Chemical Applications of Potentiometry (van Nostrand 1969)
Ionic Equilibria (Longmans 1978, and Polish trans.).
Diverse Atoms (Oxford University Press, 1998) for undergraduates.

for adult general readers:
Introducing Chemistry (Penguin 1975, and Spanish trans., with reprint).
Colour (Penguin 1983, and Princeton University Press, reprinted in Princeton Science Library Series).
Fire (Oxford University Press 1993, Dover Reprint and German trans.).

for pre-teen readers:
H_2O (1970); Metals (1971); and *Air (1973)* (Oxford University Press).

for readers of modern Greek:
Κάτι θα Γίνει (Thessaloniki, 2002), a cheerful collection of her travel experiences in Greece.

Printed in the United Kingdom
by Lightning Source UK Ltd.
129187UK00001B/407/A